Learning Google Analytics
Creating Business Impact and Driving Insights

Mark Edmondson

Beijing · Boston · Farnham · Sebastopol · Tokyo

Learning Google Analytics

by Mark Edmondson

Published by O'Reilly Media, Inc., 1005 Gravenstein Highway North, Sebastopol, CA 95472.

O'Reilly books may be purchased for educational, business, or sales promotional use. Online editions are also available for most titles (*http://oreilly.com*). For more information, contact our corporate/institutional sales department: 800-998-9938 or *corporate@oreilly.com*.

Acquisitions Editor: Andy Kwan	**Indexer:** Sue Klefstad
Development Editor: Melissa Potter	**Interior Designer:** David Futato
Production Editor: Kate Galloway	**Cover Designer:** Karen Montgomery
Copyeditor: Stephanie English	**Illustrator:** Kate Dullea
Proofreader: Piper Editorial Consulting, LLC	

November 2022:   First Edition

Revision History for the First Edition

2022-11-10:   First Release

See *http://oreilly.com/catalog/errata.csp?isbn=9781098113087* for release details.

978-1-098-11308-7

[LSI]

Table of Contents

Preface

GA4 is the biggest evolution yet of the most popular digital marketing tool used on the web, Google Analytics. BuiltWith.com estimates that around 72% of the top 10,000 websites use Google Analytics (*https://oreil.ly/MunnR*), and all those websites will be looking at upgrading from the legacy Universal Analytics to GA4 in the next couple of years. Due to GA4's new data model, the latest iteration of Google Analytics will not be compatible with its predecessors, unlike past upgrades such as from Urchin to Universal Analytics. The older systems will eventually be sunsetted, so it's realistic to say that in a few years' time GA4 could become the most popular analytics solution on the planet.

GA4 offers a new digital marketing paradigm: moving analytics tools beyond reporting what has happened toward influencing what will happen via data activation. Data activation is about making a positive effect on your website so you can see real business impact with your analytics. The trend for digital marketing over the last few years has been toward making faster decisions to help justify the cost of your website, app, or social media activity. As ecommerce booms, digital analytics have become more critical to ensure budgets are allocated correctly in a highly competitive arena.

Since GA4's predecessors, Urchin and Universal Analytics, were launched in 2005, the internet has changed to incorporate mobile apps, IoT, machine learning, privacy initiatives, and new business models—all of which require an evolution in how data is processed. GA4 incorporates features to support these new data streams and prepares you for the future of digital marketing.

Alongside its many native integrations such as Google Ads, Google Optimize, and Campaign Manager in the Google Marketing Suite, GA4's expanded usage of the Google Cloud Platform and Firebase means digital marketers now have the capabilities to build almost any data flow imaginable and scale it to a billion users. Learning to coordinate these parts enables digital marketers to more easily use their analysis to create data applications based on the same data sources, achieving quicker and more visible results for their own website.

These new opportunities require learning skills that may be unfamiliar to traditional digital marketers, so this book aims to help bring you up to speed and help your GA4 implementation fulfill its potential. We will demonstrate common use cases for GA4 data activation and provide step-by-step instructions on how to implement them, as well as introduce ideas and concepts to help you build your own bespoke applications.

I hope to give inspiration for those wishing to create their own data activation projects. Code examples will be included to help provide some templates, as well as introductions to various cloud components such as data storage, data modeling, APIs, and serverless functions to help you assess what technologies you may want to enable.

By the end of this book, you will be able to understand the following:

- What use cases GA4 integrations can enable
- What skills and resources are needed
- What capabilities third-party technology needs to fulfill
- How the Google Cloud integrates with GA4
- What data capture is necessary by GA4 to enable use cases
- The process of designing data flows from strategy to data storage, modeling, and activation
- How to respect user privacy choices and why it's important to do so

I think this is the most exciting era for working in digital analytics, simply because the potential of what you can do now is almost limitless. The cloud has made possible what was impossible for individuals or smaller companies to do even 10 years ago, and that revolution means I truly feel you're limited only by how much ambition you have. If this book can help inspire even one person to realize that ambition, it will have been a worthwhile venture.

Who This Book Is For

If you're reading this book, you're likely a digital marketer with some digital analytics background. Perhaps you're working in an agency or within a digital marketing department, such as for an ecommerce brand or a web publisher. You may be looking to justify upgrading to GA4 from Universal Analytics or have already made the switch and are now looking to make use of its advanced features. This book aims to both inspire nontechnical readers with what is possible and give enough practical information that technically minded readers can implement the use cases within the book and use the building blocks to create their own bespoke integrations.

The book aims to educate you on the features of GA4's integrations beyond the basics that you've picked up with your one to two years' experience in digital marketing. You're probably comfortable with implementing tags on the website and/or reading basic GA reports. More technical users may be using Google APIs and have some JavaScript/Python/R/SQL knowledge as well as some cloud experience.

This book is not an exhaustive roundup of GA4 features. Instead, this book focuses on what you can do today to extract business value out of your GA4 implementation, using the Google Cloud Platform to facilitate it.

Conventions Used in This Book

The following typographical conventions are used in this book:

Italic

Indicates new terms, URLs, email addresses, filenames, and file extensions.

`Constant width`

Used for program listings, as well as within paragraphs to refer to program elements such as variable or function names, databases, data types, environment variables, statements, and keywords.

`Constant width italic`

Shows text that should be replaced with user-supplied values or by values determined by context.

This element signifies a tip or suggestion.

This element signifies a general note.

This element indicates a warning or caution.

Using Code Examples

Supplemental material (code examples, exercises, etc.) is available for download at *https://github.com/MarkEdmondson1234/code-examples*.

If you have a technical question or a problem using the code examples, please send email to *bookquestions@oreilly.com*.

This book is here to help you get your job done. In general, if example code is offered with this book, you may use it in your programs and documentation. You do not need to contact us for permission unless you're reproducing a significant portion of the code. For example, writing a program that uses several chunks of code from this book does not require permission. Selling or distributing examples from O'Reilly books does require permission. Answering a question by citing this book and quoting example code does not require permission. Incorporating a significant amount of example code from this book into your product's documentation does require permission.

We appreciate, but generally do not require, attribution. An attribution usually includes the title, author, publisher, and ISBN. For example: "*Learning Google Analytics* by Mark Edmondson (O'Reilly). Copyright 2023 Mark Edmondson, 978-1-098-11308-7."

If you feel your use of code examples falls outside fair use or the permission given above, feel free to contact us at *permissions@oreilly.com*.

O'Reilly Online Learning

 For more than 40 years, *O'Reilly Media* has provided technology and business training, knowledge, and insight to help companies succeed.

Our unique network of experts and innovators share their knowledge and expertise through books, articles, and our online learning platform. O'Reilly's online learning platform gives you on-demand access to live training courses, in-depth learning paths, interactive coding environments, and a vast collection of text and video from O'Reilly and 200+ other publishers. For more information, visit *http://oreilly.com*.

How to Contact Us

Please address comments and questions concerning this book to the publisher:

O'Reilly Media, Inc.
1005 Gravenstein Highway North
Sebastopol, CA 95472
800-998-9938 (in the United States or Canada)
707-829-0515 (international or local)
707-829-0104 (fax)

We have a web page for this book, where we list errata, examples, and any additional information. You can access this page at *https://oreil.ly/learning-google-analytics*.

Email *bookquestions@oreilly.com* to comment or ask technical questions about this book.

For news and information about our books and courses, visit *https://oreilly.com*.

Find us on LinkedIn: *https://linkedin.com/company/oreilly-media*.

Follow us on Twitter: *https://twitter.com/oreillymedia*.

Watch us on YouTube: *https://youtube.com/oreillymedia*.

Acknowledgments

I'd like to thank Sanne for her encouragement and faith in me, and Rose for my most amazing daughter.

IIH Nordic has been instrumental in helping me write this book—many thanks to Steen, Henrik, and Robert for their support.

The #measure community has provided me with all of the inspiration; their ideas gave me something to write about. In particular, I'd like to thank Simo for his kindness over the years.

Thank you also to the technical reviewers who provided valuable feedback: Darshan Patole, Denis Golubovskyi, Melinda Schiera, and Justin Beasley.

## CHAPTER 1

The New Google Analytics 4

This chapter introduces the new Google Analytics 4 (GA4) and explores why it was developed. We'll see where Google felt its predecessor, Universal Analytics, was lacking and how GA4 means to strengthen those areas with the foundation of a new data model.

We'll also look at how the Google Cloud Platform (GCP) integration with GA4 enhances its functionality and get a first look at the use cases that will help illustrate the new capabilities of GA4 and get you started with your own data projects.

Introducing GA4

Google Analytics 4 was released out of beta and introduced as the new Google Analytics in early 2021. Its beta name "App+Web" was replaced with Google Analytics 4.

The key differences between GA4 and Universal Analytics highlighted in GA4's announcement post (*https://oreil.ly/kj6TL*) were its machine learning capabilities, unified data schema across web and mobile, and privacy-centric design.

Google had been planning the release of GA4 for many years before its public announcement. After its release, Google Analytics became the most popular web analytics system, yet in 2021 its design still reflected the design goals of the previous 15 years. Although the platform has been enhanced over the years by the dedicated Google Analytics team, there were some modern challenges that were more difficult to solve: users were asking for single customer views for web and mobile apps rather than needing to send data to two separate properties, Google Cloud was a leader in machine learning technologies yet machine learning was not simple to integrate with the GA data model, and user privacy was a growing concern that required tighter control on where analytics data flowed.

When it was first launched in 2005, Google Analytics disrupted the analytics industry by offering a full-featured free version of what had previously been available only in paid enterprise products. Recognizing that the more webmasters knew about their traffic, the more likely they were to invest in AdWords (now Google Ads), Google Analytics was a win-win investment that gave everyone access to the voice of their users as they browsed their website.

By 2020, the analytics landscape was much different. Competitor analytics products were launched with simpler data models that could work across data sources and were more suited to machine learning and privacy (an essential user feature). You could use the cloud to make an analytics system more open, giving more control to analytics professionals. Competing analytics solutions could even be run on Google's own cloud infrastructure, which changed the economics of build or buy. The ideal analytics solutions would have sensible defaults for those looking for quick start-up but would be more customizable and scalable to satisfy the more adventurous customer's needs.

The Unification of Mobile and Web Analytics

While its previous name of "App+Web" was replaced with GA4 at launch, the discarded name was more representative of why GA4 was different.

Up until it was sunset in late 2019, Google Analytics for mobile apps (Android/iOS) had its own separate analytics system distinct from web analytics. These software development kits (SDKs) used a different data model that was more suited to app analytics, where concepts like page views, sessions, and users all meant slightly different things, which meant they couldn't be easily compared to the web figures. Users who visited both app and web were usually not linked.

GA4's data model follows a customizable, event-only structure that was being adopted by mobile apps. Universal Analytics placed limitations on when data could be combined, known as data scoping, which meant that marketers needed to think about how their data fit within scopes such as user, session, or events. These were predetermined by Google, so you were forced to adopt its data model. With GA4's event-only approach, you have more flexibility to determine how you want your data to look.

When the old Google Analytics for mobile SDKs sunset in 2019, Google encouraged users to instead move over to the Firebase SDKs. Firebase had been developed as a complete mobile developer experience for iOS and Android with an integrated mobile SDK for creating mobile apps from the ground up, now including web analytics. The new GA4 represented an additional data stream on top: the new web stream. Having iOS, Android, and web streams all using the same system means we now have a truly connected way to measure digital analytics across all those sources.

Firebase and BigQuery—First Steps into the Cloud

For many marketers, GA4 is their first introduction to the new cloud products that are integral to the operation of GA4: Firebase and BigQuery.

Firebase and BigQuery are both products within the GCP, a broad service Google offers for all manner of cloud services. This book focuses on those products that are part of its data analytics cloud offerings, but be aware that these are just a subset of the whole cloud platform.

Firebase is a broad mobile development framework that now includes Google Analytics. Mobile developers also use it to give serverless power to the mobile apps with useful features such as remote config to change the code of deployed apps without republishing to the app store, machine learning APIs such as predictive modeling, authentication, mobile alerting, and Google advertisement integrations. Firebase is a subset of GCP services that are in some cases a rebrand of the underlying GCP product—for example, Firebase Cloud Functions are the same as GCP Cloud Functions.

BigQuery can be considered one of the gems of GCP; it's recognized as one of its most compelling products compared with the equivalent running on other cloud providers. BigQuery is an SQL database tailor-made for analytics workloads, and it was one of the first serverless databases available. It includes innovations such as a pricing model that stores data cheaply while charging on demand for queries and a lightning-fast query engine running on Dremel that offers in some cases 100x speed-ups compared with MySQL. GA360 users may already be familiar with it as one of its features was to export raw, unsampled data to BigQuery—but only if you bought a GA360 license (this was my introduction to the cloud!). GA4 BigQuery exports will be available to all, which is exciting because BigQuery itself is a gateway to the rest of GCP. BigQuery features heavily in this book.

GA4 Deployment

This book is not an exhaustive guide on GA4 implementation; a better place for that would be the resources outlined in Chapter 10. However, the book does cover common configurations that will give the whole picture, from data collection to business value.

There are essentially three ways to configure capturing data from websites: `gtag.js`, `analytics.js`, or *Google Tag Manager* (GTM). In almost all cases, I would recommend implementing them through GTM, which you can read more about in Chapter 3. The reasons for that are flexibility and the ability to decouple the dataLayer work from the analytics configuration, which will minimize the amount of development work needed within the website HTML. Developer resources will be most effective implementing a tidy dataLayer for your GTM since this will cover all your tracking needs, not just GA4 or Google tags. Any additional changes to your tracking configuration can then be done within GTM's web interface without needing to involve precious development time again for each minor edit.

With the introduction of GTM Server Side (SS), the configurations possible can also include direct integrations with Google Cloud and backend systems along with modifications of the HTTP call's requests and responses, giving you the ultimate flexibility.

Universal Analytics Versus GA4

GA4 is said to be an evolution of its predecessor, Universal Analytics (nicknamed GA3 since GA4's release), but how is it actually different?

One of the first questions people have when hearing about GA4 is "How is this different enough for me to want to change? Why should I go through the bother of retooling, retraining, and relearning a system that has worked fine for the last 15 years?" This is a key question, and this section examines why.

A dedicated Google help topic (*https://oreil.ly/G0ePW*) also covers this question.

A new data model

The first big change is in the data model itself, covered later in "The GA4 Data Model" on page 6.

Universal Analytics was very much focused on website metrics where concepts such as users, sessions, and page views were more easily defined; however, these concepts were more tricky to define for other data sources such as mobile apps and server hits. It often meant that workarounds had to be incorporated or some metrics ignored in the reports when the data came from certain sources. It also meant that some metrics didn't work well together or were impossible to query.

GA4 moves away from an imposed data schema to something that is much freer: now everything is an event. This flexibility lets you define your own metrics more easily, but for users who don't want to get to that level of detail, they also provide default auto event types to give you some of the familiar metrics.

This also means that it's now possible to automatically collect some data that had to be configured separately before, such as link clicks, so the GA4 implementations should take less experience to implement correctly, helping to lower the barrier of entry for new digital analytics users. Specialist knowledge such as the difference between a session metric and a hit metric will be less critical.

A more flexible approach to metrics

GA4 events can be modified (*https://oreil.ly/rtmxb*) after they have been sent. This lets you correct tracking errors or standardize events ("sale" versus "transaction") without needing to modify the tracking scripts—much easier to action.

When creating custom definitions for your own events, there is no predefined schema you need to remember. Create your event with optional parameters, and register it within the GA4 interface to start seeing that event appear in your reports.

BigQuery exports

Previously a GA360 feature, BigQuery exports will now be available even if you don't pay for the enterprise version of GA4. Firebase Analytics for mobile had this feature at launch, and because GA4 is only an addition to that, web analytics has it too.

This is a game changer because typically the most difficult part of a data project is getting access to the raw data beneath your applications in such a way that you can easily work with it. With GA4 BigQuery exports, you need to fill in only a few web forms to get that data flowing in near real time, ready for analytics using BigQuery SQL.

Because BigQuery is so integrated with the rest of GCP, this also means it has tight integrations with the rest of the GCP data stack, such as Pub/Sub, Dataflow, and Data Studio. These services allow you to pipe data directly from BigQuery, and since its APIs are open, it is a popular source or sink for many third-party services too.

This all means that the age-old problem of data silos, where the data you need is locked behind databases with differing company politics and policies, now has a route to a solution by sending it all to one destination: BigQuery. This is how you can start to link across sales and marketing or pull in useful second-party data such as weather forecasts more easily. In my experience, moving all the useful data to one place has had the most transformational effects on a client's digital maturity, as one of the most common roadblocks—"How do we get the data?"—is removed.

No sampling—everything is real time

A motivation for GA360 BigQuery exports was that it was one of the ways you could get unsampled data, and that is also now applicable to GA4. While the sampling limits are improved within the WebUI, the data underneath is always unsampled and available in real time. Should you ever need an unsampled export, it is available via BigQuery or by using the free Data API. This removes the barrier of paying for GA360 to have the data for some use cases that need high accuracy and real-time analytics data sources.

Privacy and digital analytics data

Users are rightly much more aware of the value of their data these days, and privacy has become a hot topic in the industry. There is recognition that users need a fully informed choice to consent to where their data is being used, and it is the website's responsibility to earn trust and correctly value that data. To help with this, Google Consent Mode is available to remove cookies and their stored personal identifiers so they are not available to Google Analytics until a user gives that consent. However, nonpersonal data can still be useful, and GA4 offers a way to model what your data sessions and conversions would look like if 100% of your users consent to giving their data. Since most often your new customers will be those most likely to not yet trust your website or give consent, this can be valuable information to help you improve your performance.

When is GA4 the answer?

Given the changes in GA4, following is a summary of opportunities GA4 offers over Universal Analytics to help with frequently asked questions:

- How can we integrate our digital analytics data with GCP to make our data work beyond the GA4 services (what this book is mostly about!)?
- How do we unify tracking users across all of our digital properties, including our mobile apps and website?
- How can we more easily make bespoke analytics implementations above the defaults?
- How can we access our digital analytics data to feed into our machine learning model?
- How can we respect privacy choices but still have some data on the performance of our website?

This section has talked about why you would use GA4 and its key differences from Universal Analytics. The fundamental source of these changes is how GA4 records its data in its new data model, which we'll go into more deeply in the next section.

The GA4 Data Model

The GA4 data model is what differentiates it from Universal Analytics. This new data model enables GA4 to offer its more advanced features. This section looks more deeply at the data model and how it functions.

Key elements of this data model include:

Simplicity
Everything is an event of the same type. No arbitrary relationships are imposed on the data.

Speed
Given the simpler data model, the reduced processing of events allows everything to be done in real time.

Flexibility

Events can be named anything up to your quota limit (500 by default). Parameters can be attached to each event to fine-tune its metadata.

We will now get into the weeds and explore the syntax of how the GA4 event hits are created.

Events

Events are the atomic unit of data capture in GA4. Each action a user makes on your website according to your configuration sends an event to Google's servers.

Here is just one event:

```
{"events": [{"name": "book_start"}]}
```

Simply counting the number of `"book_start"` events gives useful information, such as how many people have started the book, the average number of book reads per day, etc.

To ensure a collection of events is associated with one user, those events need a common ID. In GA4, this means also sending a `client_id`, which is a pseudonymous ID usually found within the GA4 cookie. This is commonly constructed as a random number with a timestamp attached when it was first created:

```
{"client_id":"1234567.1632724800","events": [{"name": "book_start"}]}
```

The preceding line is the minimum required data for events sent to your GA4 account.

Timestamps are usually given in Unix epoch time, or the number of seconds since midnight on January 1, 1970. For example, cookies with 1632724800 would translate to Monday, September 27, 2021, 08:39:56 CEST—the moment I am writing this sentence.

These examples are from the Measurement Protocol v2, which is one way of sending in events. The much more common way is to use the GA4 tracking scripts on your website or iOS or Android app to build and create these events. But I think it's useful to know what that script is doing.

The same event sent from a web tracker using `gtag()` would look like the following:

```
gtag('event', 'book_start')
```

The GA4 JavaScript library takes care of the cookie to supply the `client_id`, so you just need to supply your custom event name.

When using the GA4 tracking scripts, the library tries to help you avoid configuring common event types by providing automatically collected events (*https://oreil.ly/ fe6V8*). These cover useful events such as page views, video views, clicks, file downloads, and scrolls. This is already an advantage over Universal Analytics: what you previously would need to configure now comes standard with GA4. Less configuration means quicker implementations and fewer chances for bugs. To use these automatic events, you can choose which to turn on via enhanced measurement settings (*https://oreil.ly/NHRpH*).

There are also *recommended events*, which are events that you implement but that follow a recommended naming structure from Google. These are more tailored to your website and include recommendations for verticals such as travel, ecommerce, or job websites. These are also worth sticking to because future reports may rely on these naming conventions to surface new features. Generic recommended events (*https:// oreil.ly/JZo7Q*) include user logins, purchases, and sharing content.

Since these automatic and recommended events are standardized, if you do collect your own custom events, make sure not to duplicate their names to avoid clashes and confusions. Hopefully you can see the flexibility of the system in its attempt to provide standardization with sensible defaults to avoid having to reinvent the wheel for each implementation.

Custom Parameters

Event counts alone are not sufficient for a useful analytics system, however. For each event, there can be none or many parameters that give extra information around it.

For instance, a login event will give you the number of logins on your website, but you probably want to break that down by how a user logs in—with email or a social login. In that case, your recommended `login` event also suggests a `method` parameter for you to specify this:

```
gtag('event', 'login', {
  'method': 'Google'
})
```

If done with the more fundamental measurement protocol, it would look like the following:

```
{
 "client_id":"a-client-id",
 "events": [
   {"name": "login",
    "params": {
      "method": "Google"
      }
    }]
  }
```

Note that we have added an array of params with the extra information.

Ecommerce Items

A special class of custom parameters is items, which are a further nested array within the custom parameters that holds all the item information. Ecommerce usually represents the most complicated data streams because multiple items, activities, and data are associated with sales.

However, the principles are largely the same: in this case, the custom parameter is an array that holds some recommended fields such as the item_id, price, and item_brand:

```
{
  "items": [
        {
          "item_id": "SKU_12345",
          "item_name": "jeggings",
          "coupon": "SUMMER_FUN",
          "discount": 2.22,
          "affiliation": "Google Store",
          "item_brand": "Gucci",
          "item_category": "pants",
          "item_variant": "Black",
          "price": 9.99,
          "currency": "USD"
        }]
}
```

Combine this with the recommended ecommerce events such as purchase and some other parameters, and the full event payload becomes the following:

```
{
  "client_id": "a-client-id",
    "events": [{
      "name": "purchase",
      "params": {
        "affiliation": "Google Store",
        "coupon": "SUMMER_FUN",
        "currency": "USD",
        "items": [{
          "item_id": "SKU_12345",
          "item_name": "jeggings",
          "coupon": "SUMMER_FUN",
          "discount": 2.22,
          "affiliation": "Google Store",
          "item_brand": "Gucci",
          "item_category": "pants",
          "item_variant": "Black",
          "price": 9.99,
          "currency": "USD",
```

```
          "quantity": 1
        }],
        "transaction_id": "T_12345",
        "shipping": 3.33,
        "value": 12.21,
        "tax": 1.11
      }
    }]
  }
```

While the preceding code represents some of the most complex events sent to GA4, I hope you can appreciate the simplicity of the underlying model. By using only events and parameters, GA4 can be configured to capture complex interactions on your website.

User Properties

In addition to the event-level data, it's also possible to set user-level data (*https://oreil.ly/hrmQv*). This is data associated with the client_id or user_id you have on record. This could be used to set customer segment or language preferences.

 Be mindful here to respect user privacy choices. If you are adding information to a specific user, then laws such as the EU General Data Protection Regulation (GDPR) require that you get consent from the user to collect their data for your stated purpose.

Sending in user properties is much the same as sending in events, but you use the user_properties field instead, as well as any events you may want to send:

```
{
  "client_id":"a-client-id",
  "user_properties": {
    "user_type":{
      "value": "bookworm"
    }
  },
  "events": [
    {"name": "book_start",
     "params": {
       "title": "Learning Google Analytics"
    }}
  ]
}
```

Using gtag() would look like this:

```
gtag('set', 'user_properties', {
  'user_type': 'bookworm'
});
```

```
gtag('event', 'book_start', {
    'title': 'Learning Google Analytics'
});
```

In this section, we looked at how to send GA4 events in various ways, such as the measurement protocol and gtag, and the syntax of sending events with parameters and user properties. We now move on to how to process those events coming out of GA4 via its integrations with GCP.

Google Cloud Platform

GCP can now be firmly embedded within your GA4 system via its preexisting data analytics systems. It offers real-time, machine learning, scale-to-a-billion services that you pay for only when you use them, while also letting you divest the boring stuff around maintenance, security, and updates. Let your company focus on what it is expert at, and let the cloud take care of the noncore tasks. Via the cloud's pay-as-you-go payment structure, small teams can create services that previously would have taken many more workforce and IT resources.

In this section, we look at the GCP services you will most likely use when integrating with GA4, the skills and roles your team will need to take advantage of these tools, how to get started, how to manage costs, and how to select the right cloud service for you.

Relevant GCP Services

This book focuses more on the data applications services of GCP, but that is still a vast array of services that are being constantly updated. For a full overview beyond the scope of this book, I recommend *Data Science on the Google Cloud Platform* by Valliappa Lakshmanan (O'Reilly).

The following key cloud services are used in the use cases later in the book and have been essential in my general work. There are many different cloud services, and choosing the right one can be a bit bewildering when you are starting out. I suggest looking at the services highlighted here as useful ones to get started with.

We will become familiar with the following services in the book, in rough order of usefulness:

BigQuery
　　As mentioned already, BigQuery will feature heavily as both a destination and source for analytics and data workloads. It even has modeling capabilities with BigQuery ML.

Cloud Functions

The glue between services, Cloud Functions let you run small snippets of code such as Python in a serverless environment.

Pub/Sub

Pub/Sub is a message queue system that guarantees that each message is delivered at least once at a scale that can handle the entire internet being sent through its queue.

Cloud Build

Cloud Build is a continuous integration/continuous development (CI/CD) tool that lets you trigger batched Docker containers in response to GitHub pushes. It's a hidden workhorse behind several of my solutions.

Cloud Composer/Airflow

Cloud Composer/Airflow is an orchestrator that lets you reliably create complicated interdependent data flows, including scheduling.

Dataflow

Dataflow is a batch and streaming solution for real-time data that is well integrated with lots of GCP services.

Cloud Run

Cloud Run is similar to Cloud Functions but lets you run Docker containers containing any code you like.

There are usually a few ways to create what you need, and the differences may be subtle, but I recommend that you be pragmatic and get something that works first, then optimize which exact service may be better to run it later on. For instance, you may have a daily data import running on a BigQuery scheduled query but find as your needs get more complex that Cloud Composer is a better tool to coordinate the import.

All of these tools are not point and click, however. Coding is required to get them to deliver what you need, so we will go over what skills you need to deliver on their capabilities in the next section.

Coding Skills

One of the most daunting aspects of applying these integrations may be that it calls on skills that you may think only computer programmers have. Perhaps you consider yourself "nontechnical."

I used to think the same thing. I remember at the start of my career saying, "I don't know JavaScript" and waiting six weeks for time to free up for a developer to enact a five-line bit of code on a website. Once I found the time and the inclination, I started to have a go myself, making lots of mistakes along the way. I also learned that the professionals also made lots of mistakes, and the only difference was that they had the motivation to keep going. Another realization was that a lot of what I was doing in Excel was actually more complicated and more difficult to work with than if you were using a tool more suited for the job. Solving the task in Excel needed more brain power than doing it in R, for example.

So if you are inclined, I would urge you to keep going. If it's hard, it's not necessarily because you don't have the talent—these things are alien to everyone at the beginning. Coding can seem incredibly fussy in some cases, and things can go wrong if you miss a single ";". However, once you learn one area, the next is a bit easier. I started by being a power user of Excel, then learned Python and JavaScript, then fell in love with R, then had to learn to appreciate SQL and bash, and now dabble with Go. The nature of programming is that, as you learn and get better, the code you look at from six months ago will look awful. This is natural; the important thing is to be able to look back and see progress. Once you get something working, that is experience, and it slowly grows until 10 years later when you're sitting down writing a book about it.

For me, open source was also a way to sharpen my skills, since putting code out in the open and getting feedback was a multiplier on top of any experiences I had running that code. This is why I'm so grateful for any feedback I get today, in GitHub or otherwise. The code in this book will also be available on a GitHub repository to accompany the book, which I will endeavor to keep updated and free of bugs.

 By the same logic, if you read some of my code and have some feedback on how it can be done in a better way, please do get in touch! I am still always learning.

The use cases in this book include code examples that cover the following languages:

JavaScript
This is essential for all web page–based tracking involving HTML and is most commonly used for data capture via tags. It is also heavily used within GTM to create custom templates.

Python
A very popular language that is supported by a wide range of platforms, Python is useful to know as it can be considered the second-best language for everything. It has strong machine learning representation as well, although you probably won't need that unless you're working on advanced implementations.

R

Although you could get away with just using Python, R's data science community makes it the best language for data science in my opinion. Its libraries and open source community cover everything from data ingestion all the way to data activation via interactive dashboards and reports. I attribute most of my thinking on how to approach data workflows to the mindset I got from R, so it influences projects even when it's not directly used.

bash

When interacting with cloud servers, they will most likely be using Linux-based systems such as Ubuntu or Debian, which rely on bash to operate rather than a graphical interface such as Windows. It's also handy to know some command-line bash programming when dealing with very large files that aren't easily imported into other languages. gcloud and other CLIs also assume some knowledge of shell scripting, the most popular being bash.

SQL

In most cases, the raw data you're working with will be in a database, and SQL will be the best method for extracting it. SQL also introduces a way of thinking about data objects that is helpful.

While it may be possible to copy-paste your way to victory, I really recommend going through line by line and at least understanding what each section of the code is doing.

Assuming you now have some coding available, either through your own skills or your team's, we now move to how to get started on GCP and deploy your first code to the cloud.

Onboarding to GCP

GCP is a major component of Google's business, and it has flows that are completely separate from Google Analytics that you will need to learn to navigate.

You can get started for free, but the first thing to know is that for anything serious, you're going to need to add a payment card for your cloud usage. However, you may get several months of usage covered by the onboarding vouchers available.

Google's start-up page (*https://oreil.ly/9e6Hn*) will guide you through your first login.

 If you have an existing Google Cloud Project, it may still be worth creating a new one for the examples in this book to ensure they are activated with the latest versions of APIs. For example, you will most likely need to activate the Google Analytics Reporting API, Google Analytics Admin API, and Cloud Build API, and check that the BigQuery API is active by default.

Costs of the Cloud

The cloud offers limitless possibilities but comes at a cost. There are many free quotas for cloud services, but you must be mindful to keep an eye on costs because they can quickly add up. I've seen cases where a BigQuery SQL query was scheduled daily that used much more data than expected and then the user went on vacation. When they checked a few weeks later, the job had costs thousands of dollars! An even worse scenario would be accidentally publishing your sensitive authentication keys. There have been at least three times that I've seen those keys picked up by bots to start up expensive GPU-enabled Bitcoin mining machines, each costing thousands of dollars.

While the free tiers are usually adequate for experimenting and the pricing models are usually very generous, it's worth using the GCP pricing calculator (*https://oreil.ly/XOWeS*) or running limited versions of your apps first to assess the costs of production. Costs of services can heavily influence which cloud application you should use.

You should also be proactive in setting up billing alerts and protecting your authentication keys.

However, with all those warnings up front, companies have generally been surprised by how small the cloud costs are in relation to their value. Companies storing their data in BigQuery typically have bills of less than $100 a month in the beginning, and get larger bills only once they have created a good use case that provides value of many factors more. I'm rounding up for the $100 figure—in reality, it's more likely to be in the region of $5 until you have your live active use cases ready, but I tend to quote $100 to clients so they are pleasantly surprised if it's less than that!

Factors that influence cost are the amount of data being moved around, computing time, and how real-time your applications are. Cloud cost savings are usually due to being billed only once actual work is being performed, rather than paying a flat fee for the services. But that is also heavily influenced by what services you use; since there are usually many ways to tackle a specific problem, there is typically a way that would replicate how you would do it in your local environment, and a less-expensive way using the cloud-enabled serverless technologies that we talk about in "Moving Up the Serverless Pyramid" on page 15.

Moving Up the Serverless Pyramid

Truly unlocking the power of the cloud involves an evolution in thinking about how to tackle IT problems using its strengths. Companies' first step in the cloud usually involves a "life-and-shift" model where they simply replicate what they had running locally within the cloud, such as a local MySQL database replaced by a cloud server running MySQL. Another strategy is "move and improve," which involves, for example, putting your MySQL database within Google Cloud SQL, a managed instance of MySQL.

However, a "lift-and-shift" model will yield only minor benefits compared to the full potential of the cloud. For a company to achieve true digital transformation, it needs to embrace the higher metaservices built on top of the fundamentals of compute and storage, with the provision that doing so will necessarily tie you a bit more to that cloud provider's service.

The cloud companies' pitch to use these services is that you *divest* the IT resources to maintain, patch, and develop the services built and instead *invest* in using the applications built on top in a more on-demand manner. It is undoubtedly this model that has enabled me to write this book, as without cloud computing, creating your own services would be much more complicated and limit your ability to experiment with solutions. When IT resources are effectively outsourced, much smaller teams are needed to achieve results.

An example of this is BigQuery. Creating your own BigQuery service would require you to invest in having huge server farms at the ready, which cost money when idle just so they're available when you need the resources for a "big query." Using the Big-Query service for the same query, those resources are bought online as necessary, and you effectively pay for only the seconds they are running.

To illustrate this, I find the diagram of the serverless pyramid in Figure 1-1 helpful. It outlines some of the services and the trade-offs you get when you're selecting a service to run your use case.

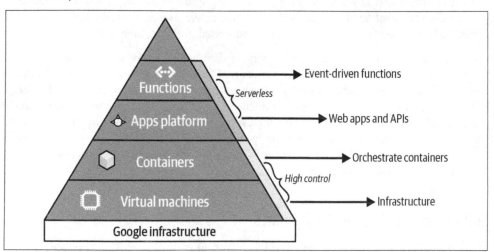

Figure 1-1. GCP pyramid hierarchy

At the bottom level, you have virtual machines and storage, which are basically cloud versions of the computers running on your desktop. If you want complete control of configuration, you can fire these up with some cloud advantages like backups, security, and patches. This layer is sometimes referred to as *infrastructure as a service* (IaaS).

At the next level, you have services that run virtual machines and storage for you but abstract it away so you only have to worry about the configurations you need. App Engine is an example of this, and this layer is sometimes referred to as *platform as a service* (PaaS).

At the level above that, you have yet another level of abstraction running on top of the equivalent PaaS. These services are usually more role-driven, so services like analytics data warehousing (BigQuery) are available. This is sometimes known as *database as a service* (DBaaS).

And even above that, you can have services that take out some of the configuration to provide even more convenience. Often you need to supply only the code you need to run or the data you want to transform. Cloud Functions is an example: you don't need to know how the function executes its code but just specify how you want it to run. This is referred to as *functions as a service* (FaaS).

With this in mind, you can judge where your application should sit. The services at the top of the pyramid typically have a greater cost per run, but if you are under a certain volume or cost of implementation, they still represent a massive cost savings. As you need to own or scale more of the infrastructure, you may consider moving down the pyramid to have more control.

The use cases in this book aim to be as high up the pyramid as possible. These services are usually the latest developments and quickest to get started with, and will give you the scale to serve you up to your first billion users.

And that is truly now within reach—a consideration when selecting your service is how much it will be used, which can include up to global Google scale. Perhaps you don't need it right now, but it is still worth considering in case you need to reengineer your application should it prove unexpectedly successful.

This is where being far up the pyramid (as detailed in Figure 1-1) is helpful, as those services usually have autoscaling provisioning. These should be limited to avoid expensive mistakes, but essentially, if you have the money, then you should expect similar performance for one thousand users as you would for one billion. Further down the hierarchy, you still have options, but you would need to be more involved in the configurations of when and where you should apply that scale.

Wrapping Up Our GCP Intro

This has been a brief whirlwind tour of why the cloud is so powerful and how its power can be applied to your GA4 implementation. We spoke of how the cloud puts resources in your hand that only a few years ago would've required a large IT team to enable, and we also talked about the concepts of serverless versus lift-and-shift models for how you may approach this. This will involve an expansion of your digital roles to include the coding languages that help enable such services, with the promise that investing in those skills will make you an overall more effective digital marketer. The majority of this book will cover how to put this into practice, with some example use cases on things you can do right now.

Introduction to Our Use Cases

This book introduces all the concepts and technologies that are relevant for GA4 integrations, but theory and planning can go only so far. The real way I learned the skills discussed in this book was by implementing applications. I made mistakes along the way, but those mistakes were often the most valuable learning experiences, because once you debug why something went wrong, you gain greater understanding of how to get it right.

To help jump-start your own journey, once all the building blocks necessary for your own applications are introduced in the following chapters, our use cases in Chapters 7, 8, and 9 are dedicated to technical use cases detailing the whole lifecycle of a GA4 data application including code examples: creating the business cases, technical requirements, and decisions about what technologies to use. If you follow along with everything in sequence, by the end you should have a working integration.

 In practice, you may accidentally skip certain steps and have to go back and carefully read what you missed. Also, by the time you implement a particular use case, the technologies may have changed slightly and need updating.

Even with a perfectly implemented example, it's unlikely that it will match exactly what your own business needs or what it should prioritize. The use cases cover my experience of common customer problems, but your own will undoubtedly be slightly different. Because you will most likely need to adapt the use cases for your own needs, it is important to understand not only what to do but also why we are doing it one way rather than another. You will then be able to adapt the process to better suit your own priorities.

Despite your individual requirements, some common themes can be tied together in relation to how to approach these projects. Chapter 2 covers a framework that every successful data integration project I've worked on has had in common. The use cases will follow this framework to give you practice in applying it. The four main areas are data ingestion, storage, modeling, and activation. However, the question the use case is asking is the main driver of all this, because if you are trying to solve a problem that isn't actually going to help your business when solved, the whole endeavor will not be as effective as you wish it to be. Finding the right problem to solve is important for your own business, which is why in Chapter 2, we'll also go through some questions you can ask to help you define it.

The practice use cases will allow you to concentrate solely on the practical work of implementation. The best way to learn will be to follow along and implement them rather than just reading through them. They can also act as a reference when you're implementing your own use cases, as you can often reuse aspects of one solution within another. For example, all the use cases in this book use GA4 as a data ingestion source. The use cases also try to use several different technologies to cover a broad range of applications.

Use Case: Predictive Purchases

The first use case in Chapter 7 is a baseline to help you get used to the overall approach that shares its structure with the more complex use cases later in the book. We'll use only one platform, GA4. The same principles still apply to the more involved use cases, but this should also show how it's possible to swap out GA4 for other applications should it better serve your needs. This case uses several of GA4's new features, including its machine learning and audience exports.

Predictive purchasing uses modeling to predict if a user will buy in the future or not. This can be used to alter the site content or advertising strategy for those users. For instance, if the probability that a user will make a purchase is above 90%, perhaps we should suppress marketing to that user because the job is already done. Conversely, if a purchase probability is under 30%, perhaps we should consider that user a lost cause. Enacting such a policy means that you can move your budget allocation to target just the 60% of users who may or may not buy. This should drive down your cost per acquisition (CPA) and potentially increase your sales revenue.

To do this, we will use GA4 to do the following:

- Collect data on the website including conversion events
- Store all the data we need
- Provide data modeling using its predictive metrics like purchase and probability
- Export to Google Ads for activation using GA4's Audiences

This process is illustrated in the simple data architecture diagram in Figure 1-2.

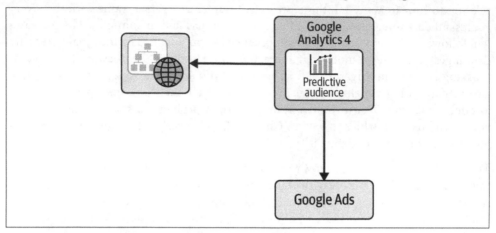

Figure 1-2. Data architecture for the predictive audiences use case

No programming will be required to enact this, and all configuration will be done within the UI.

Predictive metrics is a feature integrated within GA4 and makes direct use of Google's capabilities in machine learning to make a real difference in how your business operates. However, your website needs to meet certain criteria to qualify for using the predictive metrics feature, which puts you less in control of when the feature can be used. If you can't use predictive metrics, you may still be able to use your own data and build the model yourself and then use the Google Ads integration later. We'll cover this in the next section.

Use Case: Audience Segmentation

The audience segmentation use case in Chapter 8 shows you how to better understand the aggregate behavior of your customers. What common trends or behaviors can you pick out so you can better serve that segment? How many types of customers do you have? Do the data-driven segments you've found match the assumptions of your business?

Such segmentation projects have historically been used to help personalize marketing messages for those users. For instance, certain customers may be identified as being more likely to buy cross-sell products, so you can limit marketing messages to just those customers to reduce campaigns costs and avoid unnecessary messaging to customers who may get annoyed by them.

You can segment on many different criteria. A successful method that predates the internet is the RFM model that looks at the recency, frequency, and monetary behavior of users and segments those with similar scores in each sector. With the wealth of data available now, you can make other models with hundreds of fields. The model you choose will be largely governed by your use case's business requirements as well as its privacy considerations. Privacy is important here as it may be necessary to collect consent from users to include their data within models. If you don't collect consent, the customer may get annoyed if they're targeted.

Using this example, we would like our Google Ads costs to be more efficient. In this context, Google Ads will take on the data activation role since that is where we will send data to make a change in user behavior. Our business case is to reduce costs as well as get higher sales if we can tailor our messaging more tightly.

We'd like to use the data we have about a customer's website behavior and their purchase history to determine if we should or should not show them certain ads. To do so, we will use the following:

- GA4 and our customer relationship management (CRM) database as the data sources
- Cloud Storage and BigQuery as our data storage
- BigQuery to create our segments
- Firestore to push those segments to our GA4 users in real time
- GTM SS to enrich the GA4 data
- GA4 Audiences to push those segments to Google Ads

Interactions between these services are shown in Figure 1-3.

Along the way, we'll also ensure that privacy choices are respected and that no personal data is exported or transferred where it is not needed.

The technologies we'll use for the following services will be covered in the relevant chapters in more depth later:

- GA4 for web measurement
- A production database for user purchase history
- Imports via Cloud Storage, Pub/Sub, and Cloud Functions
- BigQuery for creating the segmentation models
- Cloud Composer to schedule the updates
- Cloud Storage, Pub/Sub, and Cloud Functions to import the segments to GA4
- GA4 to create the audiences

You'll need skills in Python and SQL as well as some configuration work within GA4, the Google Cloud console, and Google Ads. We'll also need to make sure that we're collecting the right data within GA4 so we can link the web activity with the CRM data in a privacy-compliant manner.

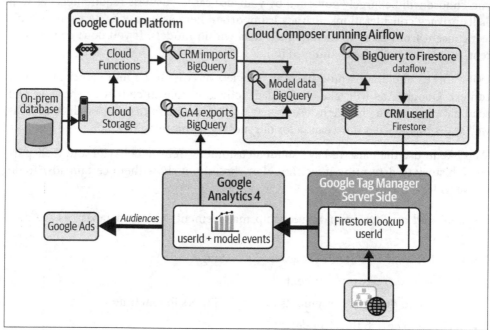

Figure 1-3. Data architecture for the user segmentation use case

Use Case: Real-Time Forecasting

The use case in Chapter 9 is about creating a real-time forecasting application. Real-time analytics is often a first ask for companies when they're getting into analytics, but it is usually deprioritized if they discover they can't react to that data stream in real time. However, if you do have that ability, it's an exciting project to work on because you can see the immediate benefits.

A good example of this use case is a publisher newsroom that is reacting to real-time events during the day when choosing which stories to publish or promote. In a company where clicks and views mean revenue, a social media viral hit can make a big business impact. To get that hit takes repeated attempts, editing and promotion on the home pages, and constant real-time feeds of trending social media topics and sentiment. The use case we detail here shows how to take that stream of web analytics data and forecast what that traffic will do based on current uptake. You can make these forecasts on GA4 Audiences that have been set up to identify different segments of your customers.

This use case will demonstrate using Docker to handle running the dashboard solution on Cloud Run, running R's Shiny web application package. A key reason to use Docker is that you can swap out code that is running within the containers with any other language, so Python, Julia, or another future data science language could be substituted. The data roles for this project include:

- Data ingestion via APIs
- Data storage within the application
- Data modeling in R
- Data activation via an R Shiny dashboard

To achieve this use case, we'll need the following:

- GA4 to collect the real-time web event stream
- Cloud Run to run the dashboard
- GA4 Audiences to have a useful segment to forecast

Figure 1-4 shows how these resources connect with one another.

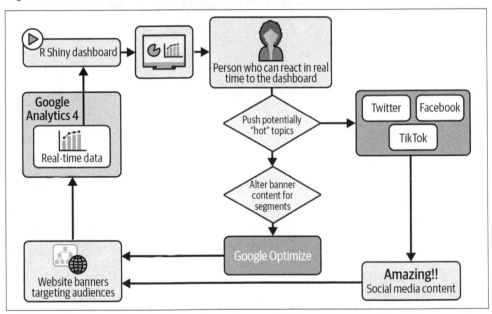

Figure 1-4. Real-time data is taken from GA4 and a forecast is created to help prioritize content for social media and on-site banners via Google Optimize

We'll use some R skills to create the real-time feeds and modeling, and we'll use some dashboard visualization skills to make the dashboard.

Summary

This chapter introduced the main ways to use GA4 to advance your digital analytics implementations. We explored why GA4 was created in the first place and how it differs and improves upon Universal Analytics with its new simpler data model. We also explored how its integration with GCP opens up your digital analytics to a whole new world of applications involving services such as Firebase and BigQuery. Although using these new cloud services requires new skills like coding, the new services of the cloud make this more accessible than in the past. Serverless architecture offerings have made it possible to abstract away a lot of the legwork in configuring and scaling up compute services. A general recommendation when starting out is to aim to use services as far up the architecture as possible to keep the barrier of entry as low as possible.

Even though the technology is now available, how to approach and best use it is a key skill that may be unfamiliar to digital marketers who haven't used the cloud before, so in Chapter 2, we set up the general framework and strategy for creating successful data analytics projects that can be repeated over many projects. We'll develop the roles of data ingestion, data collection, data modeling, and data activation from a strategic perspective, ready for the practical implementation chapters that follow.

Data Architecture and Strategy

This chapter looks at the steps you should take before you start configuring or coding anything. My perspective is mostly based on digital marketing consultancies, so it may be biased toward projects of that nature, but it also means that the processes are motivated by quality, cost, quick results, and control of resources that should resonate with other kinds of business. We'll look at how to create enthusiasm and buy-in from the relevant stakeholders, consider the pros and cons of different approaches, and then help scope out the necessary actions and requirements so we have a roadmap for how to execute. We'll also look at how to define whether a project has been successful.

Creating an Environment for Success

I wanted to at least include an overview of the nontechnical aspects of digital analytics projects, because they're so important in actually creating business results. You won't get anywhere unless you have the business itself on your side, and this is often the hardest work to justify, scope, and get approval for doing the project in the first place. We'll look at how to get your stakeholders behind a project; create an agile, use case–driven plan that demonstrates real business value; and assess if the digital maturity of your company is ready for such a project so it can benefit from it over the long term. If your company is not digitally mature, even the best data product will have no impact if, for example, the staff is unwilling to use it.

Stakeholder Buy-In

When I started trying to develop data science applications, I had a "build it and they will come" attitude. I thought that the brilliance of the application would justify its adoption, and we would simply need to show the working proof of concept for it to be enabled globally.

The reality was that people simply will not adopt highly technical applications that weren't developed in partnership with those who would benefit from them, unless you're very lucky. Nowadays, I'm very aware that the people who will work with the data need to be involved from the beginning, and they need to own the process and be aware of the business benefit so that they can comfortably explain it to their boss (and boss's boss, etc.).

The best process for this is to bring together everyone who will be involved in the same room and talk through what you want to achieve. The business value will most likely come from nontechnical participants. This is even more important when you will be using data across departments, as typically the politics of data silos will be the major stumbling block for the project. It's dangerous to think of which data application would be best from only your perspective because you may be drawn to the most technically interesting idea, not necessarily the most business-friendly one.

Another key issue relates to IT development projects in general. The digital analytics to date are initiated from companies' marketing departments and usually do not require the same best practices that IT has developed for its own systems over many years. If GA4/BigQuery is the first introduction for the company to the cloud in general, there may well be a major roadblock in getting your IT department to agree to it, especially if you plan to use your own first-party data. You have to get IT on board to have successful long-term prospects, otherwise you risk creating a "shadow IT" process that is created to navigate around restrictions your own IT department may be perceived to be blocking you with. This is not sustainable.

If it's your first project in this area, a secondary goal will be to gain trust within the company that this is a worthwhile area to invest in. The first project will introduce the cloud infrastructure for the first time and show how it is different from local development, so I recommend not starting with a big complicated project (or "spaceship" project) but rather a smaller but still useful application. Success will be showing that it all works, that costs are good, and that you have a sound basis for developing the digital maturity of the company.

A Use Case–Led Approach to Avoiding Spaceships

The use case–led approach has, in my experience, been the best way to get these projects completed. A use case gives everyone a target to aim for and answers the question of why you're doing the project. Without a use case, a technology solution can be introduced just because it has some vague promise of being beneficial. After the initial enthusiasm has waned, the project can be endangered once certain internal champions leave or the first costs come in from running it.

It's important to break down the project and restrict scope as much as possible to achieve success—I would suggest within six months. Any longer than that and you risk losing focus as people leave, and the project can turn into a big, expensive project (the "spaceship") that gathers a bad reputation for costing a lot of resources with no business results to show for it.

Being able to demonstrate quick wins is important to gain confidence in the processes, which is important when data is involved, since a loss of confidence in your processes can kill the entire analytics initiative.

"Give me the freedom of a tight brief" is a phrase I heard early on in my career, and it has stuck with me. It's much nicer to work on a project with very defined goals and actions. Don't be afraid to push nonnecessary features into a phase 2 after the initial work has been done, and be mindful of scope-creep as you work on the project. Ideally, technical requirements for the project should be fully referenced and be able to be checked off at the end of the project.

Demonstrating Business Value

Tied tightly to the use case–led approach is working out the real business revenue or cost savings your project will achieve. The better you can define this, the more confident you can be in the budgets you present.

Typically, you can look to show this value in a variety of ways:

- If automation is involved, look at how many hours per month are being used at the moment by employees and work out an average hourly cost saving once the solution is in place.

- If you're looking at increasing key metrics, pick metrics that are tied as closely as possible to revenue or cost savings. I've used the "improving page speed" metric in the past, but for many in the business, it was too far away from actual business value. "Improving total conversions" is much better because you can then multiply that by the average goal value to give you an incremental uplift figure.

- Cost savings are typically used to keep the same budget but make it more efficient, but it depends on what phase a business is in. Young companies focused on growth are normally completely unconcerned with saving costs, whereas long-established businesses with a declining market share may be looking at only costs.

Once you have some kind of cash value assigned to your use case, you can then determine what value your solution will bring. It may be that after this assessment, you realize that your solution costs too much, in which case you have saved yourself a lot of time and effort that can be directed to other more worthwhile projects.

Assessing Digital Maturity

Another key factor is that the use cases you work on must be achievable according to the companies' digital maturity at that time. It's no use pointing to the top of the mountain as a goal when you're down in the foothills without even a good pair of climbing boots.

Likewise, promising advanced real-time machine learning projects to companies that are currently using bounce rate as a key performance indicator (KPI) may get polite nods but rarely turn into actual projects. You'll need to ask lots of questions and conduct assessments to see what the next steps are in a company's journey. However, keep that mountaintop in mind as inspiration for why they should look at improving, which opens the door to creating a digital maturity roadmap over many years.

Prioritizing Your Use Cases

We now move to deciding which of your ideas you would like to work on. The process of prioritization allows you to select the projects with criteria such as the amount of resources it will require or the expected revenue impact.

Here are some questions to help you prioritize what will be on your digital marketing roadmap:

- What are the key data sources you need for your work goals?
- What are your main channels for data activation?
- What do you wish you could do with your data that you can't do now?
- What data do you think you should be able to use but can't?
- What technologies are you using for your current data work?
- What are your key business KPIs?

When working with clients, we come up with a short list of brainstormed use cases from all stakeholders and then rate each one on business impact and estimated time to execution. We then look to prioritize the quick-to-market, high-impact scoring ideas.

Technical Requirements

Once all of the stakeholders are enthusiastic about the project, we can start my favorite bit: creating the scope and technical requirements for execution. This will lay out the roadmap for how completing the project will be fulfilled and will go into as much technical detail as necessary (the plans to date may have been more high-level). All data projects have the following four elements that help you break down the phases of work:

Data ingestion
> Determining how the data will arrive, most likely in a raw state

Data storage
> Deciding how the data will be stored and be available via joins, transformations, and aggregations

Data modeling
> Turning the raw data into something useful

Data activation
> Taking the useful data and pushing it to a system that will make a business impact

Roles are helpful as the actual technologies that provide them are agnostic—most clouds will provide services that can be swapped out and matched. For example, if you were implementing a use case that used BigQuery but wanted to replicate it within Azure or AWS instead of Google Cloud, you could replace the role of data storage from BigQuery with another cloud provider's alternative (Snowflake, Azure Synapse Analytics, or Redshift perhaps).

Since this is a book about GA4 in particular, our use cases will always involve it. Its most common role will be as a data source, but it has features that actually fulfill all the other roles as well. If you use GA4 data import features such as custom data imports or Measurement Protocol, then it can fulfill a data storage role. If you use its predictive metrics, then you're using its data modeling capabilities. Then perhaps you export those metrics via Audiences for data activation. This book will also look at extending beyond those features via GA4's integrations to show its flexibility and power.

To help set your expectations, it's common for new practitioners to think that the data modeling role will take up the majority of the time within the project. In reality, it's likely to take the least time! Data modeling implementation time is usually dwarfed by the data preparation. As a rough rule of thumb, I expect the time spent to roughly fall as follows:

- Data ingestion: 20%
- Data storage: 50%
- Data modeling: 10%
- Data activation: 20%

The Right Tool for the Right Job

You will often come across tools that specialize in one part of the data flow journey but can do everything. Be cautious using those tools outside of their specialties! An example is data visualization tools that can import and transform as well as visualize data. These tools are convenient for simple data sources, but they will struggle as soon as you get to more complicated data flows, and you'll waste time trying to make them work. Leverage a tool's strengths and use other tool's strengths for the other parts of the data flow. BigQuery is a much better place for data storage and transformation, so you can use that to export out to Data Studio for visualization, which has data transformation abilities suitable for light work but not heavy-duty joins or aggregations.

Data Ingestion

Let's start with the first step in your data's journey: gathering it from the various sources available to you. In data ingestion, you collect raw data from where it is generated, such as website interactions, social media activity, or email clicks. We'll discuss how to treat data ingestion for our use cases with GA4 and GCP in Chapter 3.

The way you ingest data is typically tied to who owns or controls it:

First-party data

First-party data is your own private data. Your web analytics and internal sales or marketing systems are all in this class of data. Your own digital maturity and choice of data systems will be the major factors in how easy it is to use this data. It is quite common that the quality of the data makes it impossible to use, so a pre-project task may be cleaning it up so it's usable. An example of this could be using campaign tags or customer relationship management (CRM) database cleanups. GA4 data fits within this bracket. In many cases, for digital data it is usually easiest to take advantage of GA4's collection APIs to send data to it as the first destination, such as with custom events, dataLayer pushes, or with the Measurement Protocol. However, Google stipulates that no personally identifiable information (PII) should be sent to GA4, which means you'll be looking at streaming any PII data directly from your own systems. How easy it is to export or integrate your first-party data is increasingly becoming a factor in determining what systems to invest in. Legacy systems at a company are rarely replaced if they're functional, but the most common reason I've seen for frustration is that they are closed, walled gardens that don't allow you to really own your data because you can't extract and use it in other systems.

Second-party data

Second-party dat is another company's first-party data—an example would be Google Search Console's impression data of your SEO keywords. You would typically have a deal with that company, and the data may be provided via an API or data export. It can be useful to enhance your first-party data without having to share your own data with someone else. This data is commonly accessed via API calls to the service, or perhaps with FTP exports. In that case, you'll need to look at how you host your code that fetches the data. In some cases, like the BigQuery Transfer service, this may only require a form to be filled in by the right user. You can often use a SaaS solution for linking data, such as Supermetrics, Fivetran, or StitchData. In other cases, you build up the API calls yourself and then run them on a schedule—I typically use a combination of Cloud Scheduler, Cloud Function, Cloud Run, or Cloud Composer.

Third-party data

Third-party data is generally aggregate data from various data sources. Weather data or benchmarks are typically in this class. This kind of data can really add some context to your own and can be collected either at source when collecting other data, such as calling a weather API when a user visits to collect whether the sun is shining outside their window, or after you've collected data via an API import on a scheduled basis, like that described for second-party data.

No Personally Identifiable Information (PII) in GA4

To stress this point, this includes any data sent accidentally into GA4, as well as any sent intentionally. Usual culprits include URLs with email address included within them from form submissions, search boxes where users accidentally enter personal data, and so on. GA4 can and has closed down accounts in the past where PII data has been collected, so it's worth taking your time to ensure you don't carry this risk.

Once you've identified and how to import your data, you'll need somewhere to store it. It's time to start considering your data storage options.

Data Storage

All data is originally stored somewhere, but for your data application, you will need to decide if you can stick with its origin or if you'll need to move it to another system you have control over. We'll discuss how treat data storage for our use cases with GA4 and GCP in Chapter 4. For our use cases, the answer to this question will typically be BigQuery, since it offers many technical capabilities that are useful for data applications:

- The cost of storage is minimal
- It can accept real-time or batched data
- You can run analytical queries across terabytes of data and return results within a reasonable amount of time
- It integrates well with other systems

BigQuery is not quite suitable for every application, even though it's suitable for a wide range of use cases. If you're looking for subsecond results for a lookup query (say you're looking up a user ID and need that user's attributes), then BigQuery is not the right tool for the job. It could be part of your data application data flow to transfer data from BigQuery to Firestore so that it's available in a quicker-access format.

How do you decide which data storage solution you should use, even when not on GCP or using GA4? The goal of this section is to help you with that decision.

Your data storage needs to consider these questions:

Is your data structured or unstructured?
> Is your data in a form that can be kept within a database (CSV or JSON, for example), or is it in a form that can't be queried easily, such as images, videos, binary files, sound waves, or CSV/JSON without a set schema? One nice feature these days is that you can run machine learning over some of these forms to turn them into structured data, for example, by tagging image files.

Will you need to run analytics over your data?
> Analytics workloads favor performing calculations quickly, while nonanalytics data, such as website serving, favor fast individual-record access. A way of thinking about it is if the database used columns or rows to store data: columns are faster for SUMs and COUNTs, and rows are faster at returning one individual record.

Will you need to update with transactional-style data?
> Data updates for finance transactions, for example, may update a user's bank balance thousands of times an hour, whereas loan decisions need bulk updating only once a week. ACID compliance may be necessary.[1]

Do you need low latency in your results?
> If you need results in subsecond timing, an analytics database may not be the ideal choice.

---

1 ACID: atomicity, consistency, isolation, and durability.

Will you be integrating your data with mobile SDKs?
Google has the dedicated Firebase suite for mobile data, which lets you integrate other mobile-friendly services.

How much data will you be using?
Knowing if you're in the MB, TB, or PB range can affect your options.

How well will it integrate with your data ingestion/modeling/activation needs?
It's possible that a data storage will fulfill all your other needs, but if it's in the wrong location or not able to work with your other steps (or expensive to do so), then it may not be suitable. This could be the case if you're, for example, trying to import data from multiple clouds of locations for one application. Most clouds can create applications, but it gets expensive if you have to import/export out of one to another.

If you can answer these questions, then Figure 2-1 can help you select which tool from the GCP suite may be suitable.

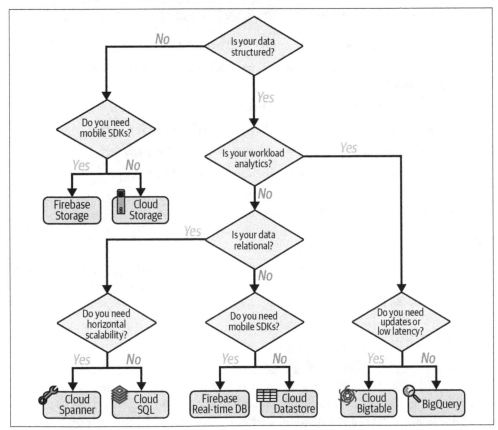

Figure 2-1. A flowchart decision tree for selecting the correct GCP storage option

Now we see why BigQuery is the answer for most of our use cases on GCP and GA4. I'll answer the questions in the figure with some typical analytics workflows:

- Is your data structured or unstructured? **Most analytics data is structured.**
- Will you need to run analytics over your data? **Yes!**
- Will you need transactional inserts? **No.**
- Do you need low latency in results? **Not for analytics workflows.**
- Will you be integrating your data with mobile SDKs? **No.**
- How much data will you be using? **All ranges; it doesn't matter.**
- How well will it integrate with your data ingestion/modeling/activation needs? **GA4 has a native integration with BigQuery.**

The answers to these questions tell you why BigQuery is a good choice. If your answers are different, you may end up with a different solution.

However, for some data activation scenarios, we arrive at a different solution:

- Is your data structured or unstructured? **Structured.**
- Will you need to run analytics over your data? **No—it's already been done in the modeling stage.**
- Will you need transactional inserts? **Perhaps.**
- Do you need low latency in results? **Yes—this will be live for users.**
- Will you be integrating your data with mobile SDKs? **Perhaps.**
- How much data will you be using? **Usually under TB range.**
- How well will it integrate with your data ingestion/modeling/activation needs? **I need a real-time API with fast response time.**

This leads to Firestore.

Once you have the data flowing into the data storage solution of choice, you can start to think about the shape of that data. This is where we begin to create value, informing the data model that will solve your use cases.

Data Modeling

Data modeling is the process of taking the data in your data storage, via the data ingestion phase, and modifying it so you can use it in your use cases. Modification covers filtering, aggregating, running statistics over it, or machine learning. The data modeling phase is where the magic happens in most projects and will often be the most bespoke. Ideally, you should spend most of your specialized resources, such as data scientists' time, here. We go into detail on the use cases in this book in Chapter 5.

Modeling includes a wide range of activities and can be as simple as providing a clean aggregate table or as complex as a real-time deep learning neural network. In all cases, the aim is to turn your raw data into gold, or, more typically, a nice flat table that can be used by your data activation channel.

Model Performance Versus Business Value

Let's first consider how much performance your model needs to have. At first, we may think the performance needs to be as accurate as possible, but this may not actually be the case.

The first key concern is to define your "good enough" metric on model performance. Naturally, your data science team will aim for the highest scores possible, but the laws of diminishing returns show that 95% may be twice as difficult to reach as 80%, and 99% could be even 10 times more difficult. Does your use case require 99% accuracy, even if it prolongs the project by a year?

Lak Lakshmanan gave a key example of this in his blog post "Choosing Between Tensorflow/Keras, BigQuery ML and AutoML Natural Language for Text Classification" (*https://oreil.ly/rPpcY*).

Lak is an accomplished data scientist in the Google Cloud team and author of *Data Science on the Google Cloud Platform* (O'Reilly). He showed that choosing between the three machine learning methods had the performance versus resource requirements as seen in Table 2-1.

Table 2-1. Accuracy versus performance versus resource requirements for various machine learning methods—adapted from research by Lak Lakshmanan (https://oreil.ly/rPpcY)

Model type	How to do it	Time to do it	Accuracy	Cloud cost
Keras trained on Cloud ML Engine	Coding Python	1 week to 1 month	Low to extremely high depending on your ML skill	Medium to high
BigQuery ML	SQL in BigQuery	About 1 hour	Moderate to high	Low
AutoML	Premade model	About 1 day	High	Medium

Given those options, it's worthwhile when scoping your use case to try and specify what performance is necessary. It may be that higher accuracy will mean higher profits, but if that can be quantified, it can then help you allocate how long your data science team should work on the models.

In many cases, it's worth getting a working model up and running quickly as a baseline, then spending the rest of your time improving its performance. If your model is successfully launched, it may be that a future revisit of the project will allocate more resources to a more accurate model.

Principle of Least Movement (of Data)

The complication of the data modeling section of the project is influenced by how much data needs to be piped to various places. Because of how web analytics data is captured, you will most likely be using only structured data for your model phase, and it's also likely to be an SQL database. Implementing complicated statistical or machine learning models in SQL is an obscure art, and your data scientist will most likely prefer to work in a more dedicated data science language such as Python, R, or Julia.

However, you should start weighing the pros and cons of moving data around to help fulfill this aim. A general principle is that you should move as little data around as possible and move only what is absolutely necessary. Following this guideline will help you avoid expensive bills and data privacy concerns, and it can enforce data cleaning from your sources so that your data scientists don't spend more time cleaning data.

Raw Data Inputs to Informational Outputs

In essence, a tight brief for your data modelers will give them a list of the data schema they will receive and the expected data format that should come out of the other end of their modeling process. In between, there may be joins, aggregates, statistics, and machine learning with neural nets or otherwise, but the machinery of the model will be an input and output dataset.

In real projects, modelers often quickly discover inconsistencies or errors within your data when they get strange results. This can be a side benefit in and of itself, since you can then feed back to the data sources and clean them in an iterative process.

Specifying your data activation channel will allow you to tightly scope the format of the data that will come out of the process. The activation channel may need to be a certain shape or in a certain system to activate—for instance, it may need to consume the data via an API or a CSV import.

Helping Your Data Scientists/Modelers

Other than a tight brief, key tasks to make your job, or your data scientists' jobs, easier allows you to spend more time on the issue and less time on the administration or data cleaning work around the data. Having a good, clean working environment also enables you to work freely on datasets without waiting for jobs to run or authorization to be given.

A good brief to your data scientist will include the following elements:

- The expected format of the data inputs alongside a detailed data catalog on what each data point represents
- The desired output metrics and/or dimensions in the modeled data
- A rough threshold on the success metric for the project (e.g., we need predictions to be more than 80% accurate to start seeing business value)
- How often new model predictions or updates will be needed
- A deadline for when the first models should be in QA
- An explanation of where the model will be deployed and the expected benefits
- If the predictions are needed in real time or via a batched process

Naturally, the best people to ask what will make their lives easier are the people working on the problem, but the preceding list will hopefully help you keep a checklist of concerns you or your resources need to address to work at optimal efficiency.

Setting Model KPIs

When scoping out machine learning modeling in particular, there are key questions about what you will use to measure success. A common example is using accuracy for unbalanced datasets such as predicting a conversion rate. Since conversion rates are typically in the 1% to 10% range, you will get a 90% to 99% accurate model just predicting that everyone will not convert! Being careful to choose the correct way to measure your model's performance is something a data scientist will be used to doing, but it's also why it's important to know the context in which the model outcomes will be used. For the previous example, recall would be a better measure, which would be the ratio of predicted conversions by the number of observed conversions.

Once the model is in production, it will generally decay over time. This is only natural as data evolves. Because of this, you should also look at setting thresholds of your model KPIs for deciding when to look at either retraining on new data, or if that doesn't work, perhaps revisiting the entire model with a new approach.

Final Location of Modeling

Once the model is created, you need to decide how your data activation will actually access it for its predictions. There are several new products that will help you "productionize" your models, which we'll explore in more depth in Chapter 5.

The key issue is getting your new data to connect with your model so that it can output its predictions or information. Although the data to train models may be large, the actual data to trigger results will usually be quite small—perhaps just a user ID or the page visited. The general trends for using your models are as follows:

Create your model where your data sits
> Databases have become more sophisticated, and many now allow you to actually create the model within them. No data movement is required between training and production workflows. BigQuery ML is an example.

Upload your model to where your data sits
> The output of your model may be an executable or binary that you can upload to where your database sits. It needs to be specifically supported by the database. BigQuery ML's Tensorflow import feature is an example.

Bring your data to your model
> The model is hosted somewhere and you upload your data to it to output your predictions. Google's AutoML services are an example.

Develop an API to access your model
> Develop an API that will return the model results when pinged with the data it needs. Machine learning APIs such as the speech-to-text API are an example. An advantage is that it can interact with anything that can talk with HTTP.

Once you are this far into the project, you should now have data ingesting in your storage solution, and created a dataset to answer your use case as well as a way to communicate it. In the next section, we describe how you justify the hard work you have put into creating your model by creating measurable business value.

Data Activation

Last but certainly not least is data activation. Data activation is so key that it should be decided within the use case's initial scope, whereas the other steps can be worked out after the scope is given. We'll go into detail about this in Chapter 6.

In this section, we'll consider different data activation possibilities related to digital marketing since GA4 will always be a source of the insights ready for activation.

Maybe It's Not a Dashboard

When working on data projects, data activation is often an afterthought that doesn't go beyond "Let's make a dashboard!" As someone who has gone through the process of creating dashboards only to find six months later that no one is even logging in, I caution you to never take it for granted that a dashboard is the best way to deliver your hard work.

My issue with dashboards is that the creators often assume that their job is finished and the viewers of that dashboard will always act upon the data viewed. We expect that the metrics and trends we show in the dashboard will promote some moment of epiphany for its viewers and that they will then rush out to enact, demonstrating the business value. If this is the desired outcome, then the dashboard should be anchored in the business, training, and workshops to help deliver its promise. This work is an ongoing task, and the dashboard should evolve given the business needs.

While I've retreated a little from my "Never dashboard!" stance, I do believe they should be involved in only the first steps of true data activation. The projects I've seen make the most impact are those where the data modeling's information has directly altered a digital marketing channel's behavior.

Take a look at the other end of your marketing stack: perhaps you have a marketing automation tool, customer data platform (CDP), or CRM sending emails that will accept integrations from your modeling. For GA4 in particular, Audiences is a data activation channel because those can be exported to your paid media channels or Google Optimize within the Google Marketing Suite. If you can link your data modeling to that activity more directly, you're much more likely to be able to demonstrate a killer use case with measurable outcomes.

Interaction with Your End Users

As we are focusing on digital marketing, the levers we can pull to make an impact all lie within the digital marketing channels. Following are the main channels with some suggestions on how your data could affect them:

Organic search and SEO
> Keyword research, matching content to queries, generation of landing page content, encouraging click-through

Paid search
> Keyword research, quality score optimization, response to trends, audience segmentation

Email
> Audience segmentation, personalization, content research

Owned media content such as your website
> Conversion rate optimization, page load experience, personalization

Social Media
> Catching trends, personalization, content research

Display advertising
> Quality assessments of placements, segmentation

In addition to serving your customers, you may also be able to help your own colleagues and internal stakeholders do their jobs more efficiently. Common avenues for this may include:

Dashboards
> Offer decision support for employees by providing information based on your data flows

Email
> Send employees useful personalized emails with your data insights included within

Automation
> Remove repetitive tasks so employees can spend their time on something more productive

Human resources (HR)
> Assess when employees need help, such as when they spend a lot of time in process bottlenecks

Stock levels
> Optimize when to order products based on forecasts of demand provided by your marketing activity

Once your data modeling is activated, you should be able to relate back to your original aims of the use case revenue and assess its impact. We have one more consideration though, which has come to the forefront in recent years: user privacy.

User Privacy

User privacy can no longer be ignored in any solutions working with data. I'm used to working within the European Union's General Data Protection Directive (GDPR) and ePrivacy legal backgrounds, and now these standards are starting to be adopted globally. Demonstrating user privacy can now be considered a competitive advantage, so the solutions you create need to be able to demonstrate that you can be trusted while at the same time generate useful results for the user if they do give you permission.

In general, the principles introduced by GDPR in the EU aren't aimed at restricting data applications of data but at protecting a citizen's dignity. As the value of a person's data increases, you must the aims of the algorithms that are deciding people's fate without their knowledge, especially if they have unwittingly given up their data for free, not for their own benefit but for corporate profit.

Other regions around the world are following the same route. Similar legislation was enacted in China and Brazil in 2020. The USA doesn't have federal-level protection, but states such as California have introduced privacy protection acts that overlap with GDPR, and other states are looking to follow suit.

A key need for user privacy is knowing the different types of data associated with a user, which differ slightly by region in legal treatment. Because my experience is within the EU, you may want to look at specifics for your own region, although the categories should generally be applicable to all locales.

Anonymous data
>Anonymous data cannot be used to re-identify a user in combination with any of the information you've gathered. This includes data that could be linked or joined to narrow down users, such as postal codes, which on their own are not identifiable, but if you link them with demographics, such as age and gender, you could possibly identify an individual. If a motivated hacker broke into all your systems, they should not be able to reconstruct a user from any data you have. A motivated hacker is a test of your data security, but a more likely day-to-day scenario is that your company leaks that data by accident by unintentionally exposing your database to the public or publishing a secret authentication key.

Pseudonymous data
>This is an ID associated with a user that when combined with other data will reveal more personal data about them. A typical example is a user ID that could be linked with a database detailing a user's name, address, and phone number. If a motivated hacker has access to the ID plus your internal systems, they could identify a user.

PII (personally identifiable information)
>This is data that directly identifies a user, such as their name, email, or credit card number. A motivated hacker would need to access only the PII data to identify a user. This also includes implicitly collected data, such as an IP address.

When designing your data applications, consider what data you actually need from a user. Anonymous data may be sufficient to provide context-based segmentation rather than individual user behavior tied to an ID. This drastically changes the amount of data you can use if you're respecting user choices and may give you a better-performing model and reduce legal risks.

Once you know what type of data you can potentially gather, how can you check that the data you are receiving are compliant? We'll discuss this in the next section.

Respecting User Privacy Choices

Knowing how you use someone's data and for what purpose they have agreed to giving it is important. Some practitioners get caught up with the technology of the systems rather than their intent; for example, if a user has not consented to have their personal data tracked using cookies, the user still shouldn't be tracked if you use a cookie-replacement technology such as a browser's localStorage. The spirit of the law needs to be respected.

When gathering consent, it's typical to gather permission for several types of usage. These types are typically broken down into necessary, statistics, and marketing roles. For instance, if you gain a user's consent for statistics, you shouldn't use their data for marketing.

When gathering consent to use PII or pseudonymous data, you will need to include that consent and when/how it was given within your datasets to keep track of users' decisions. Users could withdraw permission in the future, so you will need to have permission dates to update your records accordingly.

Privacy by Design

If possible, it's better to use anonymous data where you can because a lot of legal complications will simply not apply.

If PII data is unavoidable, then look at using pseudonymous data instead. This is encouraged under GDPR and the California Consumer Privacy Act (CCPA). With pseudonymous data, the ID is used instead of a user's name or email to give the user some level of protection if a data breach occurs—if the company has secured the lookup table of the linked user ID and their personal information.

Pseudonymous IDs also mean that it's much easier to respect user choices on deletions or data requests of portability. In those cases, you can update the central PII database and the pseudonymous ID will simply stop working without you needing to follow the trail of data through many systems to delete user data.

If you're importing PII data, the user privacy measures may already be set up. In that case, it may be advantageous to set up data expiration within, say, 30 days, meaning that if data imports stop, all cloud data will be removed within the legal time period under GDPR. If the permissions in the source data are updated, your import will then eventually reflect that change, and you won't risk including data that should be deleted in your import.

We have now covered a lot of the high-level considerations when devising a strategy for your GA4 data application project. The next section will outline some helpful tools that I use in most of my projects, so you can become familiar with them.

Helpful Tools

This section looks at other tools outside of GA4 or GCP that I consider to be essential for smooth operations. Not using them and running projects is possible, but the tools highlighted here are will make things a lot easier for you in the long run.

gcloud

gcloud is the command-line tool that enables you to do everything (and more!) you can do in the Google Cloud web console via the command line and bash programming. I consider it an essential part of your kit to help you work with the GCP, since you always have a route to automation. You don't even need to install it if you don't want to—it's supplied within the cloud shell available in your browser when you log in to your GCP web console.

The WebUI in GCP is actually just one application of the underlying GCP APIs that every GCP service works behind. GCP takes an API-first approach that means all features are available first in the API, then in tools such as gcloud or the other SDKs.

Visit the gloud CLI overview (*https://oreil.ly/E3bOB*) to install it.

Version Control/Git

Even if you're not part of a large team, using a version control system such as Git is invaluable for smooth operation. Having an infinite "undo" on your code, documentation, and procedures is one benefit; being able to reliably duplicate work across machines is another. Setting up workflows that automatically check and deploy your applications in response to code is a huge time-saver. These benefits multiply when more than one developer maintains your code base.

By far the most popular hosted Git repository on the web is GitHub, which is a public website dedicated to using Git. Other systems are also popular, such as GitLab and

Bitbucket, and Google has its own via Code Repositories. Pick one that integrates best with your workflow, but do pick one.

Integrated Developer Environments

When I first started, I used to develop my code using notepad or text files, and programming seemed laborious and harder than it needed to be. Integrated development environments (IDEs) are programs that are essentially glorified text editors but with many specific features that make it easier to run, test, and debug your code. I use multiple IDEs depending on what features I need my code to do: RStudio is amazing for R workflows, PyCharm is popular for Python applications, and VS Code is great for everything else because it has so many plug-ins for working with tasks such as creating SQL scripts.

Containers (Including Docker)

Another technology that I now consider essential for my day-to-day work is Docker. Docker allows you to create containers running on your computer that could even have different operating systems (Linux running on Windows, for example). Docker acts as a standard way to maintain the environment your code executes within, and that means that your code and workflows are much more agnostic about where they can be deployed.

Docker helps you create a small ZIP file that acts like a mini virtual machine: an entire OS lives within it with exactly the right dependencies you need to run your code.

Cloud companies have embraced Docker because they can easily duplicate systems on a local computer within their own systems, and several of their offerings take care of the infrastructure so you only need to worry about your code. See "Moving Up the Serverless Pyramid" on page 15.

I've seen the benefits of placing code in Docker because it enables me to quickly evolve which cloud service I can deploy to, even when it's within the same cloud such as GCP. Initially, a lot of my workloads ran on a Google Compute Engine instance. When services such as Cloud Run and Cloud Build came out, I switched to use the exact same Docker image to run the same code within those serverless environments, without needing to update my code at all.

Part of the fun of working with these new tools is keeping up-to-date with what the community is using to make their lives easier. The tools I've mentioned are a small sample of the tools available and will give you a good start. I chose these tools in particular because they also introduce new modes of thinking about projects.

Summary

In this chapter, we've discussed everything you need to consider before even touching code or configuring applications. I also introduced the framework that will govern how the rest of the chapters are structured: we'll take a deep dive into ingestion, modeling, and activation, and then use those roles together within the use case chapters. We also discussed how integral user privacy is for your applications and talked about some tools that will make your day-to-day work easier. If you're coming from a pure digital marketing background, there can be years' worth of new skills to learn and develop, which from my experience has been a worthwhile direction. Should you master all the aspects talked about in this chapter you're well set for your future career within digital analytics.

The next chapter discusses the first of the data roles: data ingestion. In particular, we'll focus on how you can configure GA4 and its new capabilities to get the data you need.

## CHAPTER 3

Data Ingestion

This chapter discusses the first stage of a data analysis project—getting the data into your systems so you can work with it. For this book, this will always include, but not be restricted to, GA4.

However, having the ability to ingest data from many sources is powerful because you can merge complementary systems, and typically the insights you glean from your data are more powerful. To help get you there, the next section details how to pull data from these different systems.

Breaking Down Data Silos

The more data sources you have, the more complicated a project gets. This occurs not only for technical reasons such as finding common join keys but also due to company politics as you involve more stakeholders who control different data sitting separately within a business organization. This is often referred to as *sitting in data silos*, where an organization may have a lot of good data but it's unconnected and in different systems so it's hard to make use of it. The politics of merging data can usually be solved only by involving the stakeholders as soon as possible, ideally when you are creating the business case to use that data in the first place.

This can feel like an impossible mountain to climb when you first start. A good way to take the first step is to make sure you're not asking for more data than you actually need. In some cases, aggregated data is plenty for you to get started, rather than the initial dream of merging every individual raw data point.

Less Is More

A common thought when looking to import data from multiple systems is to try and import everything "just in case." I would instead have clear use cases that specify what data each one will need and then import only that. If other use cases come up later, alter the imports then, but second-guessing what should be sent makes projects more complicated and often preserves technical debt that your import has a chance to reduce.

Try to import only data that you have a data specification for. It is common for old databases to have columns that were put in by colleagues who are long gone and no one knows what they do, especially as older databases typically have nondescriptive column names like XB_110 due to legacy restrictions on data labels.

Also consider the type or structure of the data in your data sources. A fresh data import is a good time to clean up date formats or ambiguities in currency formats and to remove null or nonsense records.

 There is only one correct date standard: YYYY-MM-DD. Your mission, should you choose to accept it, is to eliminate all others you may encounter! For reference, see ISO 8601 (*https://oreil.ly/JEoaW*).

When importing data, this is your first chance to really get to know your data, so scoping out the characteristics or the data schema you are using is of value in itself. Simply coordinating across the company so that everyone calls the same data points by the same name can be an early value driver!

Specifying Data Schema

While you may have the option of autodetecting schema, if you don't have a sophisticated way to evolve your schema on import, then it is always better to specify exactly what you expect to see in your imported data in production. Autodetection may be helpful in the development phase, but tightly specifying the name and type of a column means you can pick up future errors quickly. Generally, the best approach when dealing with data quality is to fail fast and correct data mistakes as soon as you see something unexpected, rather than having bad data points invisibly filter through to your production systems. This helps foster trust in your data project.

GA4 has its own schema as defined by Google, which you will learn when configuring it to capture data from your website. Let's look at how to specifically configure GA4 to make best the use of it.

GA4 Configuration

Because this is a Google Analytics book, we'll now deep-dive into GA4 configurations for data ingestion. We'll be using GA4 data in all the use cases outlined in the book, so getting to know what GA4 is capable of and the best uses for its data collection capabilities will shape the entire downstream data project.

There are many ways you can configure the events coming into your GA4 account, so the following sections will give you a brief overview of each configuration option so you know which to apply when designing your data streams.

GA4 Event Types

The key data element of GA4 is the event, and there are many configurations for how to collect them that: automatic events, enhanced measurement events, recommended events, and custom events. Google has tried to cover both with easy defaults and the ability to customize where you want to so that you can get going quickly and create a bespoke digital analytics tracking solution quickly.

Automatic events

Automatic events are GA4 events that you do not need to configure. They are sent in by default and are fundamental to the basic reports in GA4. They cover what are judged to be the most common requirements for web tracking and have a unique status of needing no configuration to enable, apart from putting the GA4 tracking script on your website. You do not need custom code to collect them.

The automatic events cover more ground than what was by default tracked in Universal Analytics. They include the usual suspects such as `page_view` and also useful events that previously you would have had to configure yourself. Now instead of adding code to track page scrolls and video plays or search result pages, they are captured automatically.

You can find a current list of what is captured in the GA4 documentation (*https://oreil.ly/EmECs*).

Enhanced measurement events

Although automatic events data collection is on by default, you can choose which of those fields you would like to see in your reports. You can simply flip a switch within your GA4 web interface to use them, which is typically done in the web stream configuration during account creation.

Once you flip that switch, the automatic events are surfaced in your reports with no code changes required.

There are also options for more advanced configuration settings, such as how a page view is triggered during single-page applications (changes in browser history events) or the search parameters you wish to use for site search. You can see this via the links in the screenshot in Figure 3-1.

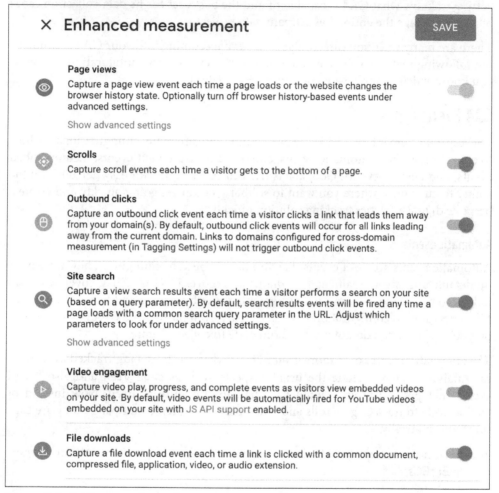

Figure 3-1. Setting up enhanced measurement configuration for a new web stream

Recommended events

Recommended events are events that are not automatically captured but have a suggested naming structure and schema that are heavily recommended you follow. They allow some customization but are not as free-form as custom events (see "Custom events" on page 51).

Google maintains documentation (*https://oreil.ly/A62Pf*) for recommended events.

Recommended events can't be collected automatically without your input on configuration because they're using data that's unique to your website, such as your ecommerce product names, but you are asked to follow a certain schema so that they will surface in the dedicated reports within GA4 in a standardized way. You're quite free to ignore those recommendations, but then you will lose functionality within the GA4 interface and APIs.

Although they are labeled only *recommended*, those recommendations are strongly suggested because otherwise some GA4 features simply won't work. Some examples are predictive measures and ecommerce, which won't function if you're not using the recommended schema for ecommerce (*https://oreil.ly/xfApC*), such as purchase, which comes with its own parameters and item syntax (*https://oreil.ly/ZjMrQ*). However, you do get a chance to modify sent-in events later using GA4's modify event feature, so this can help with accidental misconfiguration or difficulty enforcing standards across the whole organization.

Custom events

Once you have needs that go beyond the automatically collected events or the recommended events, you get to custom events.

For companies just starting on their digital measurement journey, the GA4 defaults will be great to get you started. However, as businesses become more and more reliant on their data to provide business impact, the need for bespoke customization increases, in which case you have the flexibility within GA4 to start providing your own events tailored to your more digitally mature use cases. You will need bespoke requirements unique to your business, and these could separate you from your competitors.

Each custom event you create can be made with up to 25 event parameters. This leaves a lot of room for your GA4 to be tailored to you. Once you've created the custom event, to actually see it in the reports you will need to create a custom dimension that is defined by that event within the GA4 interface.

As an example of the need for custom events, in Chapter 9 we'll talk about an online publisher looking for more meta information around their articles. The page categorization data it's looking for is not covered by automatic or recommended events.

To add the company's custom data, we need extra code configuration to capture the identified dimensions that will help in its analysis, namely the article author, category of the article, when it was published, and the amount of user interaction with it via comments or social media shares.

Example 3-1 details how this custom data may be encoded in their GA4 data collection. A simple count of the `article_read` custom event will give the total article readers, as distinct from total page views. The custom parameters are used to add the other information to this data point, with data populated from the website's backend systems and pushed into the GA4 custom event.

Example 3-1. An example of gtag() capturing data for an article_read event

```
gtag('event','article_read', {
    'author':'Mark',
    'category':'Digital Marketing'
    'published': '2021-06-29T17:56:23+01:00',
    'comments': 6,
    'shares': 50
});
```

Other default parameters will be automatically collected with the event, so you don't need to worry about duplicating those so they appear in the rest of your reports. Necessary fields such as `ga_session_id` and `page_title` are automatically collected. You can see this when you review the DebugView or the Realtime reports in the GA4 WebUI for your custom events.

Figure 3-2 shows an example from the `article_read` custom event. In the screenshot, you can see that the standard parameters are included.

There are many types of GA4 event data you can collect and many ways to do it, which you can read more about in Chapter 10. One of the most common methods for collecting GA4 event data, and the one I use the most is GTM, so we'll cover that in the next section.

Figure 3-2. Custom events and their parameters

GTM Capturing GA4 Events

Example 3-1 uses the Google Analytics native JavaScript library `gtag()` to help illustrate what data to capture, but the more typical way to capture GA4 events is by using GTM and its GA4 Event template. Rarely have I worked on a website that is not using a tag manager of some sort, since it offers wide benefits and flexibility when working with tracking tags.

GTM is a service within the GMP that complements to Google Analytics. GTM helps users control the tags that are deployed on their website in one central location, rather than having to deal with each tag individually. Digital marketers typically configure Google Analytics via GTM because it involves less back and forth with your website development team to implement changes to your GA and other tags. Using GTM means that data collection is abstracted to the GTM dataLayer, which then passes data to GA4 and any other tags you may wish to activate, such as Facebook or Google Ads.

Figure 3-3 is an example of how that configuration might look. Instead of writing JavaScript code, you can fill in a web form within the GTM interface to configure the tags, which makes the process simpler, reduces the need to know JavaScript to configure tags, ensures code standards are met, and makes data schema standards easier to meet.

Ideally, the majority of your digital marketing data captured via the website should be pushed to GTM via its dataLayer, but you also have the option of using GTM's selection of web scraping tools to lift the data off the page without coding updates to populate the dataLayer.

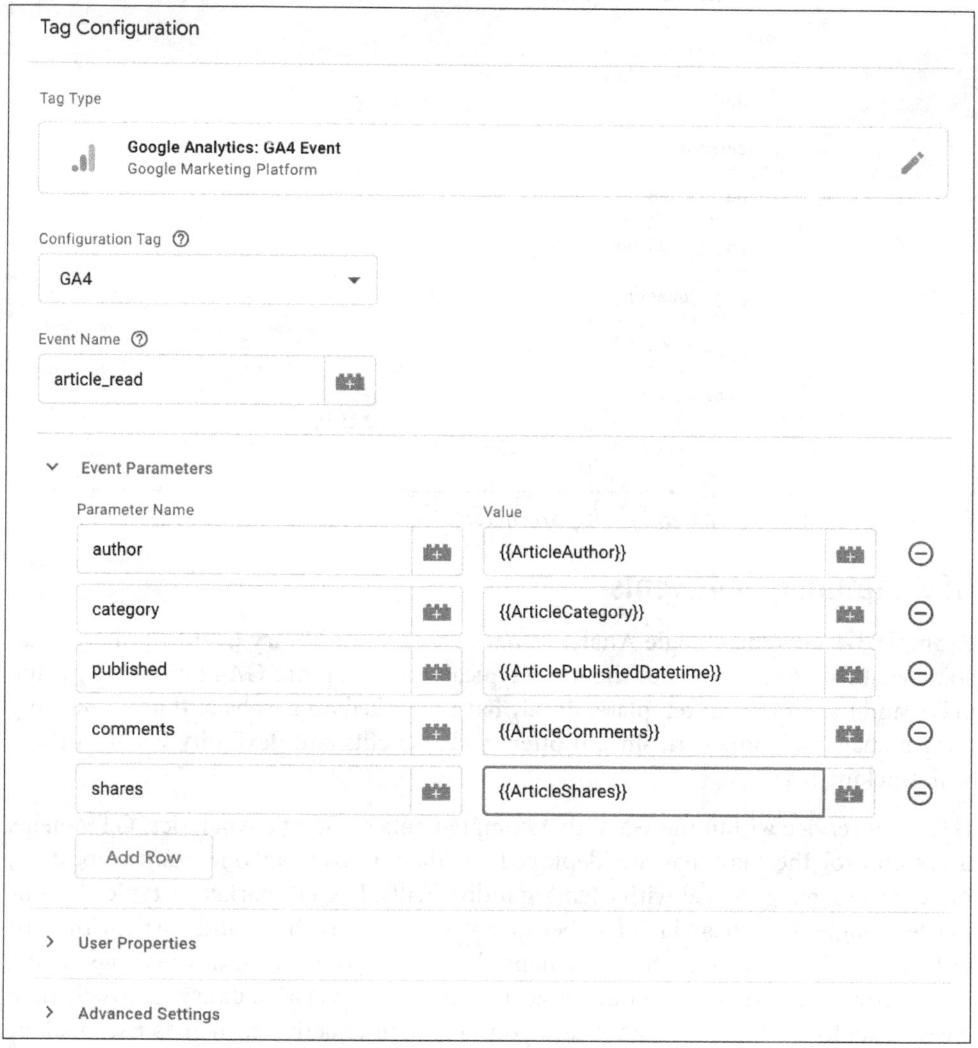

Figure 3-3. A suggested GTM configuration for sending a GA4 custom event: *article_read*

The reason the dataLayer is preferred is that it should be less prone to breakage when unexpected changes occur on your website—GTM's web scraping selectors will break if the theme or layout of the page changes. The best long-term and robust way to avoid this is to involve the web development team in your data analytics process for updates to the dataLayer, rather than relying on GTM alone. GTM should not be regarded as a way to try to circumnavigate the web development team but to make things easier for them to support your web tracking efforts.

Figure 3-4 shows an example of document object model (DOM) selection via GTM from my own blog, whose theme outputs the article publish date and time. For this particular data, the blog publish date is available via the CSS selector `.article-date > time:nth-child(2)`.

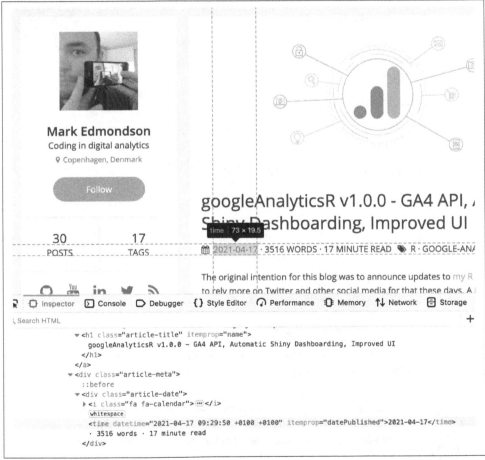

Figure 3-4. The blog post's publish date is available in the page HTML

Using GTM's DOM Element Variable, the data can be surfaced for use with the GA4 and other tags. The CSS selector code is put within the configuration, as shown in Figure 3-5.

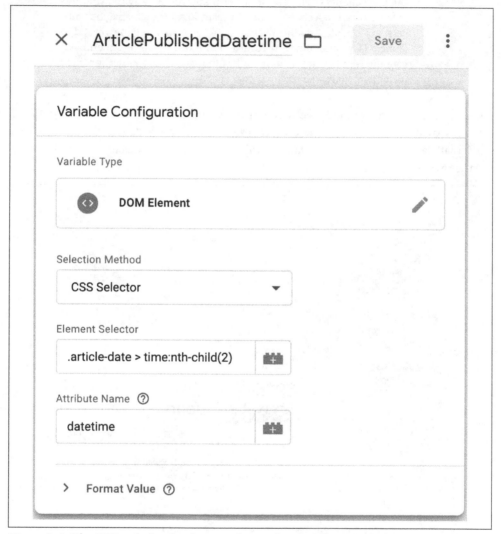

Figure 3-5. *The CSS code for the data can be used within the DOM Element Variable in GTM for use with GA4 and other tags*

 To make debugging easier, in your GA4 configuration tag in GTM, set a field called "debug_mode" = true. This will then surface the hits within GA4's DebugView found in the configuration section.

Most of the examples in this book assume you're using GTM, as it's what I use and the most popular option within the digital analytics community. We'll now look at how to configure GA4 so that you can actually see the events coming in by creating custom fields.

Custom Field Configuration

Once you've collected your events, to see them within the GA4 interface or the API, you need to configure a custom field to record the event data.

 You can read more about custom dimensions and metrics (*https:// oreil.ly/nYDWc*) in the GA4 documentation.

The full raw custom event data is available with no further configuration within GA4's BigQuery exports, so you could also replicate it there using SQL, but for use within the web interface and the Data API, you'll need to configure custom fields to tell GA4 how you want that data to be interpreted.

In our example, we'll map the `article_read` event so it creates multiple useful custom dimensions and metrics. We'll do this within the "Custom definitions" configuration screen with GA4's web interface, as seen in Figure 3-6. Here we select the `article_read` event and then choose which of its parameters will populate the custom dimensions.

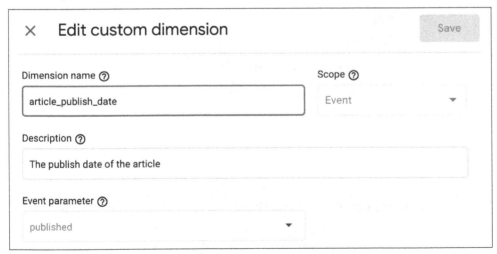

Figure 3-6. Configuring a custom dimension from the `article_read` *event*

Once done, GA4 needs 24 hours to register the event, and after that it will be available within GA4's WebUI and the Data API.

If you're collecting the custom event and mapping it correctly to a custom dimension, you should now be able to see it within your reports and your API responses. In some cases, you may not have full control of your data collection, or you may need to book developer resources each time you want to make a change. To help with that, GA4 has configuration options to help you make changes without needing to update the tracking scripts each time.

Some useful dimensions to have available are the client_id and session_id of the user. Simo Ahava has a guide on how to configure GTM (*https://oreil.ly/vFrxS*) to collect these values using his GTAG GET API tag template, which you can see in action in Figure 3-7.

Figure 3-7. From Simo's blog post, "Write Client ID and Other GTAG Fields Into dataLayer"

Modifying or Creating GA4 Events

GA4 lets you configure events after you've captured them, to help tailor and enrich your reports to your needs. The modifying events feature takes your existing event data streams and make them more useful without needing to reconfigure the data collection scripts on the website. Once configured, all future events will be processed using the rules you have set up.

Consider the `article_read` event from Example 3-1. It includes a `category` custom parameter that takes the article tags for my website, but the data is a bit messy, and multiple tags or categories are recorded that make it more difficult to analyze (see Figure 3-8).

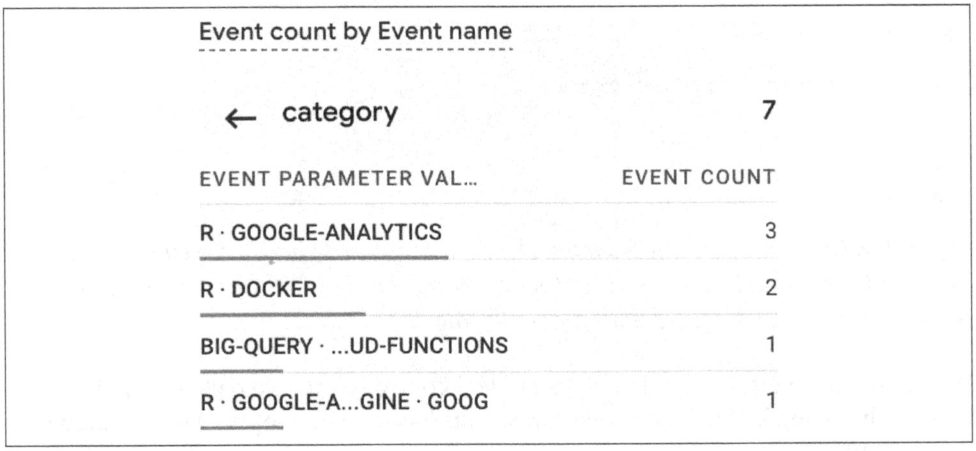

Event count by Event name	
← category	7
EVENT PARAMETER VAL...	EVENT COUNT
R · GOOGLE-ANALYTICS	3
R · DOCKER	2
BIG-QUERY · ...UD-FUNCTIONS	1
R · GOOGLE-A...GINE · GOOG	1

Figure 3-8. The `article_read` event contains a `category` parameter that has messy data due to multiple tags being recorded

To demonstrate creating custom events, we'll create some based off the `article_read` event that will fire once for each category, so, for instance, if an `article_read` event contains the categories "R" and "Google Analytics," we will also trigger an `r_viewer` and `googleanalytics_viewer` event that can be used for easier future analysis.

An example of setting up the new event using the "Create events" within the configuration UI of GA4 is shown in Figure 3-9.

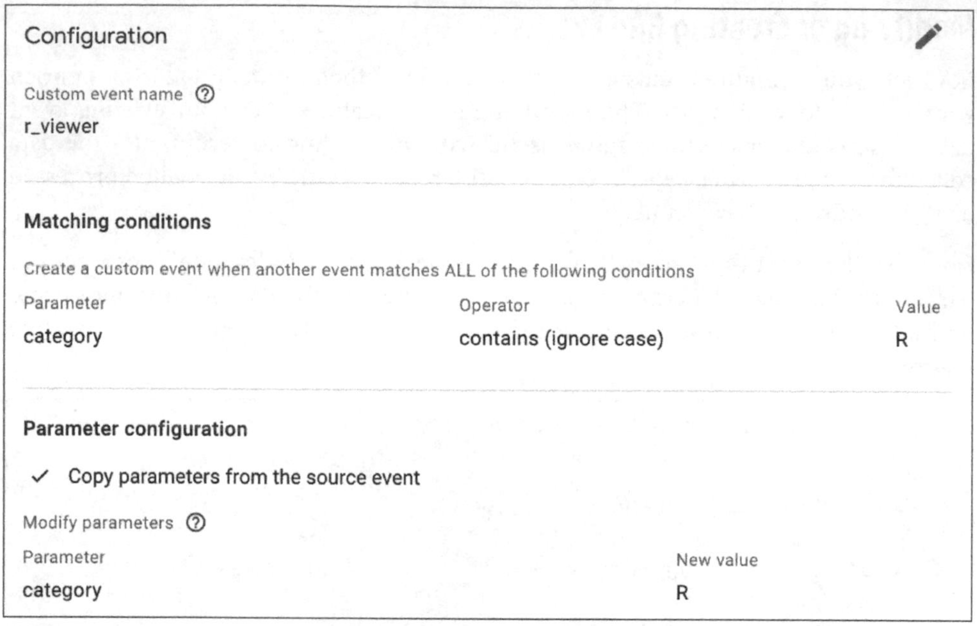

Configuration

Custom event name ⓘ
r_viewer

Matching conditions

Create a custom event when another event matches ALL of the following conditions

Parameter	Operator	Value
category	contains (ignore case)	R

Parameter configuration

✓ Copy parameters from the source event

Modify parameters ⓘ

Parameter	New value
category	R

Figure 3-9. Creating a custom event based off data captured via other events: in this case, r_viewer *is derived from the* category *parameter within the* article_read *event, selecting only those categories that contain the tag "R"*

We can make several events. In this example, I copied them across changing the criteria slightly to quickly make several more events based on my top categories, shown in Figure 3-10.

This feature is a marked improvement over Universal Analytics in which getting similar results would involve modifying the data streams as they came in via GTM, filters, or otherwise.

Events cover data you want to send in on a per-hit basis, but you may wish to persist some data for the user. This is where user properties come in.

Custom events					
Custom event name ↑	Matching conditions				
r_viewer	category	contains (ignore case)	R		>
googleanalytics_viewer	category	contains (ignore case)	google-analytics		>
docker_viewer	category	contains (ignore case)	docker		>
bigquery_viewer	category	contains (ignore case)	big-query		>
gtm_viewer	category	contains (ignore case)	google-tag-manager		>

Figure 3-10. Several created events based off of the custom category parameter from `article_read`*—joining "R" are* `"googleanalytics"`*,* `"docker"`*,* `"bigquery"`*, and* `"gtm"`

User Properties

User properties are an opportunity to add segmentation associated with your users. Different from event data, they need only be set once to be associated with that user (or more strictly speaking, the user ID associated with their cookie). The intention for this data changes more slowly than per hit or page view, for example. This is data that is more linked to a collection of hits, such as user preferences.

If you capture a user property but want to make sure it can never be used within personalized ads, you can select the "Mark as NPA" (no personalized ads) option when configuring the user property, as shown in Figure 3-11.

Figure 3-11. Marking a user property as NPA to avoid using it in targeting audiences

An example of this is to keep a record of a user's privacy consent choices: this will allow you to be confident in which users have opted in or out of your more focused targeting. For the EU, for example, the statistical consent will be distinct from a marketing or personalization consent.

Consent state can be sent once your consent management tool updates to that customer's latest choices. For instance, upon receiving marketing consent, the `gtag()` could send the following:

```
gtag('set', 'user_properties', {
  user_consent: 'marketing'
});
```

Depending on your consent solution, you can enable this in various ways. We'll work though an example using Google Consent Mode, which integrates with various Cookie Management Tools.

Google Consent Mode has various types of storage permissions, as shown in Table 3-1. For this example, we map the choices to GDPR categories. We'll assume that if you have given permission for `ad_storage`, you have also given permission for all the others. You may want to modify this for your own policies.

Table 3-1. Google Consent Mode consent types

Consent type	Description	GDPR role
`ad_storage`	Enables storage (such as cookies) related to advertising	Marketing
`analytics_storage`	Enables storage (such as cookies) related to analytics, e.g., visit duration	Statistics
`functionality_storage`	Enables storage that supports the functionality of the website or app, e.g., language settings	Necessary
`personalization_storage`	Enables storage related to personalization, e.g., video recommendations	Marketing
`security_storage`	Enables storage related to security such as authentication functionality, fraud prevention, and other user protection	Necessary

Within GTM, you can use this to create a Variable Template that will output the user consent choice.

I recommended that you do all custom JavaScript within GTM with its Templates, if possible, and not via a custom HTML or custom JavaScript variables. Templates are more optimized for caching and loading than custom HTML and have security benefits such as being able to comply with your organization's Content Security Policy (CSP).

To enable this within GTM, go to the Templates section and create a new Variable Template, and then within the templates code tab, copy across the suggested code found in Example 3-2.

Example 3-2. Creating a GTM variable for user consent

```
const isConsentGranted = require('isConsentGranted');
const log = require('logToConsole');

// the default
let consent_message = "error-notfound";

//change message according to highest consent found
if (isConsentGranted("functional_storage")){
  consent_message = "necessary";
}

if (isConsentGranted("security_storage")){
  consent_message = "necessary";
}

if (isConsentGranted("analytics_storage")){
  consent_message = "statistics";
}

if (isConsentGranted("ad_storage")){
  consent_message = "marketing";
}

if (isConsentGranted("personalization_storage")){
  consent_message = "marketing";
}

log("Consent found:", consent_message);

return consent_message;
```

We'll also need to set permissions for the template to allow it to access the consent state (see Figure 3-12).

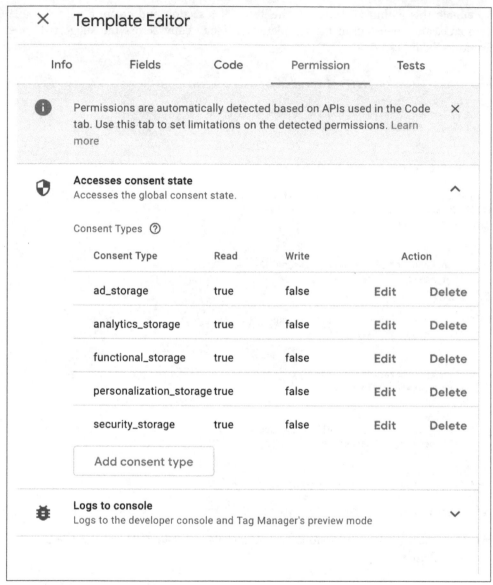

Consent Type	Read	Write	Action	
ad_storage	true	false	Edit	Delete
analytics_storage	true	false	Edit	Delete
functional_storage	true	false	Edit	Delete
personalization_storage	true	false	Edit	Delete
security_storage	true	false	Edit	Delete

Figure 3-12. Permissions for the template code in Example 3-2

Consent doesn't necessarily need to be gathered via website cookie consent tools. For example, you could give consent choices within your CRM system, or you could send a measurement protocol hit as described in "Measurement Protocol v2" on page 69.

With the variable template done, you can create a variable instance as in Figure 3-13. This will now be available for all your tags, but for our example, we'll create a GA4 tag.

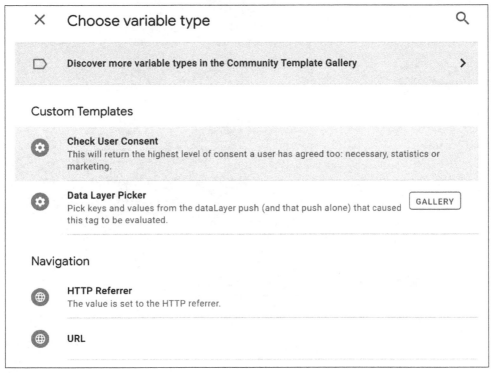

Figure 3-13. Creating a variable {{UserConsent}} from the template created in Example 3-2

The GA4 consent will be sent in with its own GA4 event that should be triggered upon receiving user consent choices or updates. Because this consent can change over time, it will be sent as an event and as a user property: the user property user _consent will reflect its current status, and the event event_consent will track when it was made (see Figure 3-14).

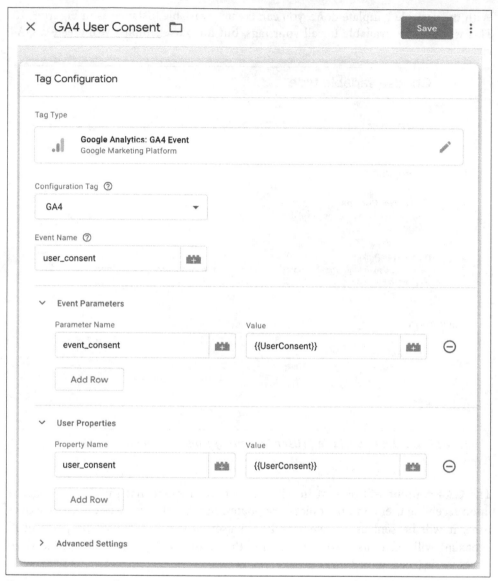

Figure 3-14. Using the consent variable from Figure 3-13 in a GA4 Event tag; the trigger should be when the Consent Tool updates

Finally, the data needs to be mapped to custom dimensions within the GA4 interface. We'll create two to reflect the latest choice and history of user consent: User Consent (Figure 3-15) and Event Consent (Figure 3-16). Here we're also determining the scope of each custom dimension, which is how long that information is meant to "stick." For the User Consent example, we want to be able to remember that preference for as long as the user is around, so we set the Scope to "User." However, we may also want to know when exactly that consent was given, so we also use an "Event Consent" with only that single event.

 When creating custom fields, the event parameter drop-down will populate with events sent up to 24 hours ago to the GA4 property. However, you can still add events manually without needing to wait for them to appear in the suggested entries.

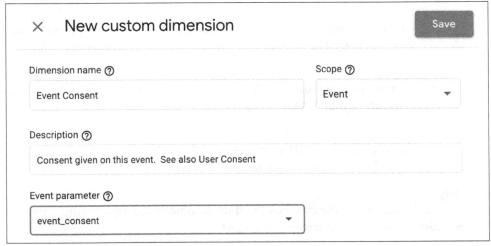

Figure 3-15. Setting up a User Consent parameter in GA4

X New custom dimension	Save

Dimension name ⑦

Event Consent

Scope ⑦

Event

Description ⑦

Consent given on this event. See also User Consent

Event parameter ⑦

event_consent

Figure 3-16. Setting up an Event Consent parameter in GA4

You can now keep track of a user's consent status as they browse the website. You can use this data to optimize when to ask for consent from a user—the first landing page is rarely the best place to ask for enhanced permissions. Gain trust from that user, then think about asking again with information about what they will get if they do allow it.

Google Signals

If you're looking to link your Audiences across GA4 to other Google Marketing Suite products, then you need to enable Google Signals in your GA4 configuration. This carries privacy implications that may not be appropriate for your company, and you should perform a privacy implication review before you enable it. You can read more about Google Signals in the documentation (*https://oreil.ly/M5y7P*).

Google Signals will link data for those users who have turned on Ads Personalization and are signed in to their Google accounts. With these extra data points, you can enable more features than before, mostly linked with using data outside of the website session that user is browsing, such as cross-device reporting, cross-device remarketing, and conversion reports. Extra data will also be available from those platforms, such as user demographics and their interests as judged by Search Ads 360.

Turning on Google Signals also has some reporting consequences. Of note is that it affects the threshold limits on data collection before it starts sampling, which may affect your data quality, and it also adjusts how an individual user is identified. The Reporting Identity documentation (*https://oreil.ly/PrrkU*) identifies several configurable ways a user may be identified, known as Identity spaces:

Google Signals
    If activated, Google will use its own available data points, such as people it can identify who have logged in via their own Google accounts and have opted in.

User ID
    You may supply your own user ID that is created via your backend systems and include them within your GA4 hits.

Device ID
    You can use a client ID or ID from the mobile device, such as recorded in a cookie. This is the closest technique to Universal Analytics.

Modeling
    If users decline any of the methods listed here, some of the gaps in your data can be deduced via modeling those sessions instead.

These identity spaces are then combined via three options in your GA4 configuration:

Blended
Looks for a user ID, and if none is present, it uses Google Signals, a device ID, or modeling as a final fallback.

Observed
Looks at the user ID, Google Signals, or device ID.

Device-based
Looks only at the device ID and ignores all other IDs.

So far, we've sent data to GA4 assuming that a user is browsing a website. But what about off-website events, like off-website transactions or subscriptions? To allow for this use case, server-to-server hits to GA4 are permitted through the Measurement Protocol (MP).

Measurement Protocol v2

The Measurement (MP) Protocol for GA4 (*https://oreil.ly/dJczG*) is a key tool for working with GA4 data ingestion. It allows data to be delivered from any location that can connect to the internet via HTTP.

The MP is intended for server-to-server communication, a slight narrowing of scope from the previous iteration, v1, for Universal Analytics, which was an open protocol that could be used to send data in from anywhere. This unfortunately led to a lot of abuse by spammers, so the new MP v2 introduces authentication so that only data from sources you trust can send data.

To help illustrate its role, Figure 3-17 from the MP documentation (*https://oreil.ly/A5IQs*) shows the interaction between the Web GA UI, the various APIs for extracting data and BigQuery, and the data ingestion.

The MP for GA4 has a slightly different scope than what was available for Universal Analytics, so we'll look at how and when you should use it in the next section.

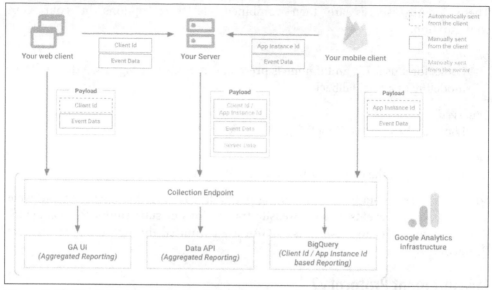

Figure 3-17. The MP sits within the data collection infrastructure and the payload in the center corresponds to MP hits, which always need to have a client ID associated

Roles for MP

The MP is intended for events that are tied to an individual user that happen away from the web environments. If you want to import batched data (such as user segments), then the Data Import (*https://oreil.ly/R6MIs*) may be a better option.

However, if a user with a known cookie ID or user ID is performing actions that would be useful to tie with the website or app activity, then you can fashion the hits so that they also appear within the GA4 reports. In all cases, you need a cookie ID or user ID so it can associate the event with the right user.

Some examples include the following:

Subscription orders
> Your eshop could have options for setting up hands-free subscriptions that repeat orders on a regular basis. Those orders would traditionally not be in GA4 since it would only track the setup or changes of the subscriptions. Using the MP, you could have your server send off the generated orders from your subscriptions so you have more accurate data on the user's lifetime revenue for optimizing your website and marketing campaigns.

Point-of-sale transactions in shops
> Use a membership number or something similar to register purchases from the HTTP point-of-sale-enabled sales register. This could allow you to measure the impact of offline to online sales and vice versa.

Off-website transactions

Calls to a call center can include sales that could be attributed to the caller's online sales. Including these sales could affect campaigns and website personalization, perhaps to surface the most pertinent content based on the products the user has purchased.

CRM updates

As a user changes their status, each update can be sent to GA4 to keep in synch with any audiences the user may be a part of (permissions, special offers, etc.).

Digital activity

There may also be activity you would like to reflect in the GA4 audiences that are exported to other activation channels such as Google Ads or Optimize. Perhaps that user receives an email or interacts with a social media post, and you would like to reflect that in your segments.

Within these roles, you have the ability to really "close the loop" on your digital analytics data by accounting not just for website activity but also for the impact of that activity on your offline channels, relating that back all the way to your marketing spend to drive traffic to your website in the first place. Achieving this is somewhat of a "holy grail" of digital marketing because it gives a more complete picture of how your marketing efforts are working than by just examining website behavior alone.

The next section will cover how to use the MP in a server environment to import that off-website data into GA4.

Server-to-server imports are one method of importing data into GA4, but you may also want to reverse that by exporting that data from GA4 into external systems for processing. This is where the GA4 APIs come into play, covered in the next section.

Exporting GA4 Data via APIs

Earlier we looked at how to send data in to GA4, but this data collection chapter wouldn't be complete without also detailing how you can import data into other external systems, where GA4 is a data source, not a data destination. Here we're typically moving to the next phase of your data collection pipeline, where GA4 has all the event data you want but can't process it in the way you need for your use case. In that instance, we look at exporting data from GA4 into external systems. The two common ways to do this use GA4's APIs and BigQuery.

GA4 data may not be the only data you need for your application, and a common workflow is to combine its data with other data from other third-party systems, usually via their own implementations of APIs. How this is achieved is usually unique to each data source.

To import data via third-party APIs, you are largely at the mercy of how good that implementation is. Although there are some standards such as OAuth2, in many cases, the format for each API will be different and bespoke. This is why it may be preferable to use a service such as Supermetrics, Fivetran, or StitchData to help import those APIs into your data warehouse.

Many APIs are also simple enough to implement yourself as a cost-saving measure, although when assessing what resources it will take, you should also include the upkeep and maintenance costs over many years.

For this book, we'll look at how to export via GA4's API in the next section, and the BigQuery exports are introduced in "BigQuery" on page 76.

GA4 has two APIs that are the descendants of the Universal Analytics equivalents: the Data API and the Admin API. The Admin API is use to set up GA4 accounts, and the Data API extracts the actual data you have collected.

 Confusingly, the old Universal Analytics API's latest version is called Reporting API v4, but this should not be used for GA4. The correct name for GA4 exports is the Data API.

Once you have enough data, you have several options for extracting via the Data API, but I most often use R and the package `googleAnalyticsR` (*https://oreil.ly/hVh5Y*) since, well, I wrote it.

GA4 Data API SDKs for Other Languages

Google supports several other languages that have SDKs you may prefer to use. See the client libraries reference (*https://oreil.ly/fWy7O*), which lists the following:

- Java
- Python
- Node.js
- .NET
- PHP
- Ruby
- Go

If that isn't enough, the underlying JSON API (*https://oreil.ly/VsKBM*) is available to make your own HTTP calls (this is what I did with the R library).

The use cases for using the Data API are slightly different from the older Universal Analytics Reporting API, since the BigQuery exports exist for all GA4 properties, not just GA360. A common reason to use the API in the first place was to help mitigate sampling, but that is not necessary because you can now extract every hit from Big-Query with no sampling. However, the Data API is a lot easier to use and is also the right price—free! The responses are also a lot quicker, so for real-time applications, the API is preferable to BigQuery exports.

The aggregations, calculated metrics, and data formats are also easier to work with, so if you're looking to quickly get started with data applications without having a complicated cloud pipeline involving BigQuery-scheduled SQL, the Data API is preferable. However, if you're looking for individual hit-level data and can handle the batched nature of the BigQuery exports, then they will be the best source.

In all cases, when using APIs you'll need to first authenticate to prove that you're permitted to read that data, which is unfortunately usually the hardest step. I'll walk you through it in the next section.

Authentication with Data API

Whichever Data API SDK you choose, you'll need to authenticate with an email that has access to the GA4 property you want to get data from. You do this via the OAuth2 process, as do all Google APIs.

> There is a more general OAuth2 library for all Google APIs available for R via `gargle` (*https://gargle.r-lib.org*) and my own `goo gleAuthR` (*https://oreil.ly/2ZqhJ*).

You can authenticate with your own email and save those authentication details locally so you don't have to repeat the authentication steps each time.

> Another option to mention is via service emails. These are emails created in the Google Cloud Console that are more suited to server-to-server applications. This is what you should use if you're scheduling a script or running it in an automated fashion so you don't expose your own personal data, and it will be more resistant if, say, you left the company but people still relied on your script.
>
> OAuth2 service emails are single-purpose accounts that you create, which you can then add to your GA4 account as users. This also provides better security because if the authentication is ever exposed, you can rotate the key in the cloud console and that key can be configured to allow access only to GA4 and not other more expensive services such as Compute Engine.

You can find out how to set up the R library via the googleAnalyticsR setup page (*https://oreil.ly/hBiA8*).

An example of how authentication occurs with googleAnalyticsR is shown in Example 3-3. When you use ga_auth() for the first time, it will ask you to create email credentials. You can then authenticate in the browser via your email. The next time you wish to authenticate, ga_auth() will give you an option to use those credentials again.

Example 3-3. Authentication with the GA4 Data API using googleAnalyticsR

```
library(googleAnalyticsR)

ga_auth()
#>The googleAnalyticsR package is requesting access to your Google account.
#> Select a pre-authorized account or enter '0' to obtain a new token. Press
#>Esc/Ctrl + C to abort.

#> 1: mark@example.com
```

In serious production work, you'll also need to create your own client key for your own Google Cloud Project, since the default Google Project googleAnalyticsR uses is shared with other users and thus subject to the same quota (around 200,000 API calls per day). The website has details on how to set this up, which you can do after you're comfortable with the initial steps to download data.

Once you are authenticated, you can start with the good stuff—seeing your data, which we'll start with via the Data API.

Running Data API Queries

For all SDKs, you'll have access to the same dimensions and metrics (*https://oreil.ly/v7KAy*), as well as any custom fields you have configured.

You also have access to the Realtime API. This has a more limited subset of the dimensions and metrics (*https://oreil.ly/BTtuB*), although it's much better than the Universal Analytics equivalent.

In all cases, you'll need to specify the GA4 propertyId you wish to query along with the date range, dimensions, and metrics you want to export.

The propertyIds for your account can be found in your WebUI, or you can query them using the Admin API, as shown in Example 3-4.

Example 3-4. Querying your GA4 properties via `ga_account_list()`*; the* `propertyId` *is the one used in the Data API calls*

```
ga_account_list("ga4")
# A tibble: 2 x 4
#  account_name      accountId property_name          propertyId
#  <chr>             <chr>     <chr>                   <chr>
#1 MarkEdmondson     47490439  GA4 Mark Blog           206670707
#2 MarkEdmondson     47490439  Another GA4 Property    250021409
```

You now have everything you need to make your first API call. In this example, we look at page views per URL. Looking through the API's allowed dimensions and metrics (also accessible via `ga_meta("data")`), you can then provide those and the `propertyId` to the `ga_data()` function, as shown in Example 3-5.

Example 3-5. Making your first Data API call, providing the `propertyId`*, dates and metrics, and dimensions*

```
ga_data(123456789,
        metrics = "screenPageViews",
        dimensions = "pagePath",
        date_range = c("2021-07-01", "2021-07-10"))
#i 2021-07-10 11:08:12 > Downloaded [ 52 ] of total [ 52 ] rows
# A tibble: 52 x 2
#    pagePath                    screenPageViews
#    <chr>                       <dbl>
# 1 /                            134
# 2 /r-on-kubernetes/            98
# 3 /gtm-serverside-cloudrun/    81
# 4 /edmondlytica/               79
# 5 /data-privacy-gtm/           73
# 6 /gtm-serverside-webhooks/    72
# 7 /shiny-cloudrun/             61
# ...


# a handy API call is one that lists your events
ga_data(
    123456789,
    metrics = c("eventCount"),
    dimensions = c("date","eventName"),
    date_range = c("2021-07-01", "2021-07-10")
)

## A tibble: 100 x 3
#    date        eventName      eventCount
#    <date>      <chr>          <dbl>
# 1 2021-07-08 page_view        239
# 2 2021-07-08 session_start    207
```

```
# 3 2021-07-09 page_view          203
# ...
```

There are many more capabilities such as creating calculated metrics, filters, and aggregations, which we won't go over here due to space constraints, but the process is similar to the one shown here. See the `googleAnalyticsR` website (*https://oreil.ly/ tqR6E*) for more applications, or the Google Data API documentation (*https://oreil.ly/ HFE6w*) or the relevant SDK's website for more details.

A big feature of GA4 is the availability of the BigQuery exports, which will cover a lot of the API use cases Universal Analytics favored. We cover these BigQuery exports in the next section.

BigQuery

BigQuery is largely regarded as the crown jewel of GCP since it was the first "server-less database" that showed astonishingly quick analytics speeds compared to the more traditional MySQL databases. It is used extensively throughout the data analytics workflows on GCP, so familiarity with it is recommended. One major benefit of GA4 over Universal Analytics is the integration with BigQuery, which allows you to reach the raw data underneath the web reports—something that was available only to enterprise premium users before via GA360. This section will go into how to make use of those GA4 exports to BigQuery.

Linking GA4 with BigQuery

The Google Analytics help article "BigQuery Export" (*https://oreil.ly/9srT3*) includes a video on how and why to link your GA4 property to BigQuery.

If you plan to keep your exports, add a Billing Account to your GCP project first to make sure you aren't using the BigQuery sandbox, which puts expiration dates on your data. Other than that, its fairly easy to use, which is one of the awesome things about GA4 having a BigQuery integration. One notable difference compared to Universal Analytics GA360's BigQuery export is that there is no historic export available, so turn on the exports sooner rather than later even if you're thinking of only using the data contained within.

See Figure 3-18 for an example made for this book: I selected EU as the location because I'm usually working for EU and GDPR jurisdiction clients and streaming to open up more real-time use cases in the future. Note that this does mean more Big-Query charges, though. I'm also choosing to put it in the same GCP project as the CRM imports needed later to make the queries a little more succinct, but that isn't a requirement since BigQuery can query over multiple projects and datasets.

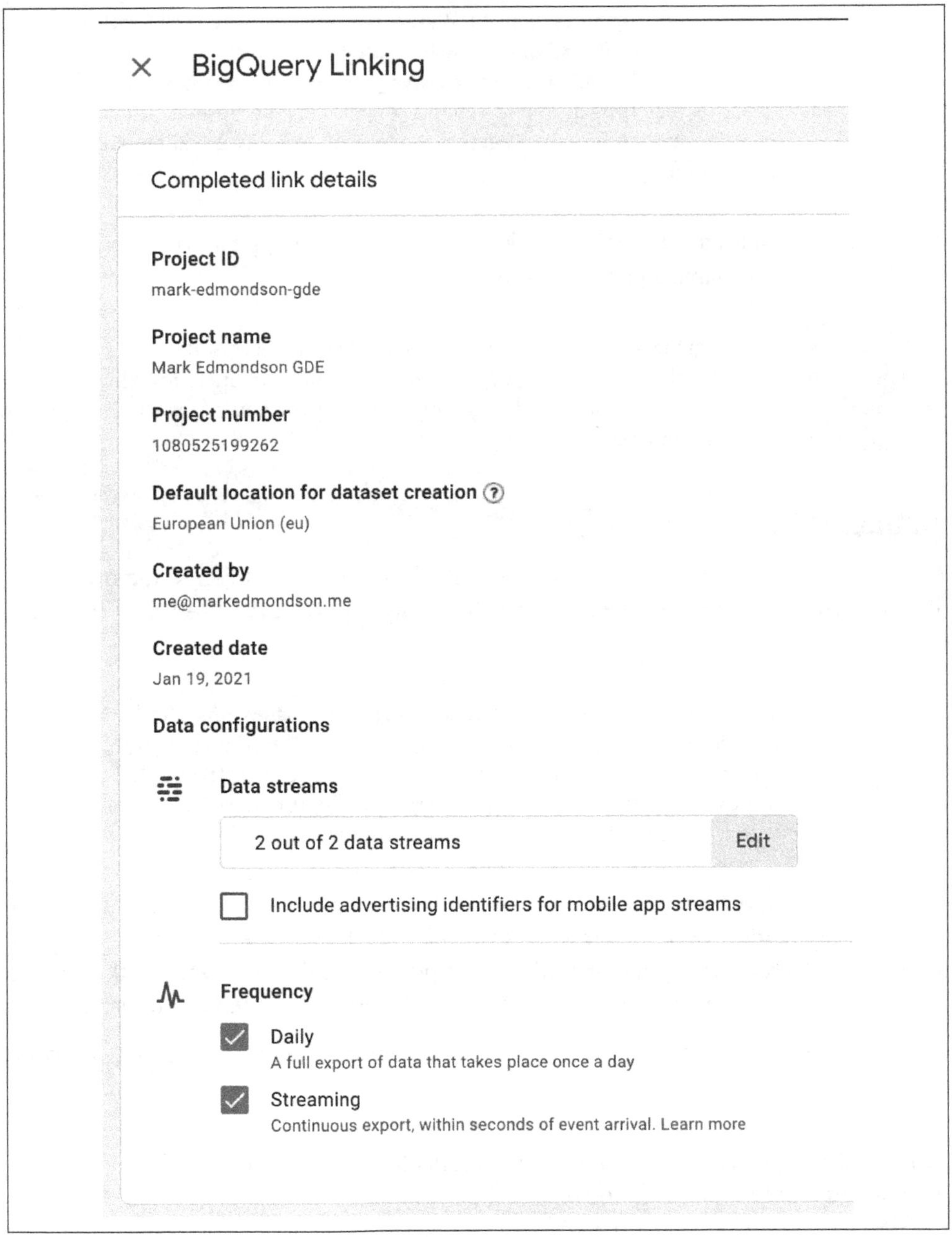

Figure 3-18. An example of a completed linking to BigQuery from the GA4 configuration screen where both Daily exports and Streaming are selected

When exporting GA4 data to BigQuery, you have the option of *Streaming* or *Daily*. Streaming will create the `events_intraday_*` tables, and Daily will create the `events_*` table. The Streaming data is more real time but less reliable than Daily because it won't include any late hits or processing delays that the Daily table will account for.

Once linked, you'll have to wait a while for the dataset to appear within BigQuery, which will have the name `analytics_{yourpropertyid}`.

The BigQuery data schema is available at this GA4 BigQuery Export schema article (*https://oreil.ly/SFxW7*). It includes the dimensions and metrics you can query via SQL so you can use it to plan your queries.

BigQuery SQL on Your GA4 Exports

The data is as granular and raw as you can get, tracking events down to the microsecond. In theory, you should be able to replicate any report in the GA4 user interface from it.

The GA4 exports use a nested data structure that is nontrivial to extract using BigQuery SQL. If this is your first exposure to SQL, this may be discouraging since the SQL involved to extract data is more complicated. Don't despair! Try out SQL on an easier dataset that uses a more traditional, *flat* data structure first.

To help you query the data, Johan van de Werken has a nice website dedicated to showing some SQL examples for the GA4 BigQuery export (*https:// www.ga4bigquery.com*), which includes examples beyond the scope of this book, including examples for many of the reports you can see within the GA4 interface.

Example 3-6 is adapted from his site and shows how you can extract all your `page_view` events.

Example 3-6. SQL to extract the `page_view` events from a GA4 BigQuery Export (adapted from Johan van de Werken)

```
SELECT
    -- event_date (the date on which the event was logged)
    parse_date('%Y%m%d',event_date) as event_date,
    -- event_timestamp (in microseconds, utc)
    timestamp_micros(event_timestamp) as event_timestamp,
    -- event_name (the name of the event)
```

```
    event_name,
    -- event_key (the event parameter's key)
    (SELECT key FROM UNNEST(event_params) WHERE key = 'page_location') as event_key,
    -- event_string_value (the string value of the event parameter)
    (SELECT value.string_value FROM UNNEST(event_params)
      WHERE key = 'page_location') as event_string_value
FROM
    -- your GA4 exports - change to your location
    `learning-ga4.analytics_250021309.events_intraday_*`
WHERE
    -- limits query to use tables that end with these dates
    _table_suffix between '20210101' and
    format_date('%Y%m%d',date_sub(current_date(), interval 0 day))
    -- limits query to only show this event
    and event_name = 'page_view'
```

If executed correctly, you should see a result similar to Figure 3-19.

Query complete (0.6 sec elapsed, 7.9 KB processed)

Job information | Results | JSON | Execution details

Row	event_date	event_timestamp	event_name	event_key	event_string_value
1	2021-07-04	2021-07-04 20:05:36.976033 UTC	page_view	page_location	http://code.markedmondson.me/
2	2021-07-05	2021-07-05 03:00:42.436612 UTC	page_view	page_location	https://code.markedmondson.me/r-on-kubernetes-serverless-shiny-r-apis-and-scheduled-scripts/
3	2021-07-05	2021-07-05 03:06:15.170026 UTC	page_view	page_location	https://code.markedmondson.me/gtm-serverside-cloudrun/
4	2021-07-05	2021-07-05 04:29:24.472675 UTC	page_view	page_location	https://code.markedmondson.me/gtm-serverside-webhooks/
5	2021-07-05	2021-07-05 05:17:04.233441 UTC	page_view	page_location	https://code.markedmondson.me/data-privacy-gtm/

Figure 3-19. An example result for running the BigQuery SQL on your GA4 exports from Example 3-6

BigQuery for Other Data Sources

A big strength of BigQuery is that it can be used for multiple data sources and facilitate breaking down data silos. Its API helps with the process, which allows integrations with most other data sources.

Data Transfer Service
: Included with BigQuery is this dedicated service that allows companies to build direct integrations. Naturally, other Google services such as YouTube and Google Ads make use of this service as a way to export their data so it appears next to your GA4 data.

  There are also many other third-party services (155 at the time of writing) that include other cloud providers (*https://oreil.ly/ve57a*) such as AWS and Azure and digital marketing services such as Facebook, LinkedIn, and Instagram. Most non-Google transfers do carry an extra charge, but for some services such as Google Ads, you can import them for free. The Google Ads import in particular is one of the easiest ways to access Google Ads data.

Other transfer services

There is also a range of services that enable transfers into BigQuery for you from thousands of other third-party services. Most of these have coded-up connections using the BigQuery API. They certainly offer a route into getting started with your data lake and should offer peace of mind in keeping up with API changes with the providers.

We've talked about how to query your own data, but often a good source of value is to merge your own data with other publicly available data, a lot of which is already available within BigQuery.

Public BigQuery Datasets

Once you have a BigQuery account, you have the option of querying any BigQuery dataset you have access to and making your own data public, should you wish. This enables generic datasets via the Public Data (*https://oreil.ly/VwZCn*) service, which offers paid and free data that may be useful for your business.

Examples include weather data, crime data, real estate listings, demographics, and country calling codes, all of which could be a significant variable in user behavior (in some cases, far more than a marketing campaign!).

GTM Server Side

GTM is already well represented in this book since it's the recommended way to configure your GA4 tags. It can also be used as an activation facilitator because you can pass data back to a user's browser that can be used to alter content. This ability is further enhanced with GTM Server Side (SS), which allows deeper integration with GCP. For instance, what if you wanted to customize your BigQuery export or have it in more real time? GTM SS can help by letting you control direct writing to BigQuery, which we'll demonstrate in the next section.

Server-side write to BigQuery

If you deploy GTM SS, you gain more control of the HTTP requests your tags make and receive. This on its own allows for enhanced data governance and privacy control.

GTM SS also lets you do more since the code it runs is not publicly available, meaning you can run authenticated API read/writes.

A first use case for this is allowing direct writes to your BigQuery datasets, something that couldn't be done in GTM standard since you would expose your authentication keys.

A GTM SS Tag template is shown in Example 3-7. It shows the code for a template that uses the stripped-down version of JavaScript you're allowed to use within GTM SS. This limited sandbox is in place to ensure that malicious or broken code can't be accidentally introduced to your server.

Example 3-7. Code for a template in GTM SS for writing the event data to BigQuery

```
const BigQuery = require('BigQuery');
const getAllEventData = require('getAllEventData');
const log = require("logToConsole");
const JSON = require("JSON");
const getTimestampMillis = require("getTimestampMillis");

const connection = {
  'projectId': data.projectId,
  'datasetId': data.datasetId,
  'tableId': data.tableId,
};

let writeData = getAllEventData();

writeData.timestamp = getTimestampMillis();

const rows = [writeData];
log(rows);

const options = {
  'ignoreUnknownValues': true,
  'skipInvalidRows': false,
};

BigQuery.insert(
  connection,
  rows,
  options,
  data.gtmOnSuccess,
  (err) => {
    log("BigQuery insert error: ", JSON.stringify(err));
    data.gtmOnFailure();
  }
);
```

The template sets up some fields that you will need to fill in once you create an instance of the template. The template fields should input the projectId, datasetId, and tableId of the BigQuery table you've set up already, and that table should match the schema of the events you're sending in. This will differ based on the exact events you're sending, and anything not specified in the schema will be silently dropped, so you'll probably want to examine the GTM SS preview logs for the exact data you want to surface in BigQuery. To help you get started, the BigQuery schema in Example 3-8 holds data for most GA4 `page_view` type events.

Example 3-8. Example schema you can use when setting up BigQuery tables to receive the GTM SS data

```
timestamp:TIMESTAMP,event_name:STRING,engagement_time_msec:INTEGER,
engagement_time_msec:INTEGER,debug_mode:STRING,screen_resolution:STRING,
language:STRING,client_id:STRING,page_location:STRING,page_referrer:STRING,
page_title:STRING,ga_session_id:STRING,ga_session_number:STRING,
ip_override:STRING,user_agent:STRING
```

All in all, BigQuery is likely to be a big part of your GA4 data flow due to its native export and other capabilities, so getting comfortable with using it is a great next step if you're interested in taking your digital analysis beyond the GA4 web interface. It also opens you up to the many other integrations on Google Cloud and elsewhere, and I would attribute it as the start of my cloud journey. If you want to really come to grips with it, I suggest *Google BigQuery: The Definitive Guide* by Valliappa Lakshmanan and Jordan Tigani (O'Reilly) as a good companion to this book.

There are circumstances, though, when BigQuery is not suitable, such as for more unstructured data that doesn't easily fit within its columnar structure. In those instances, we turn to the next most common ingestion product on GCP: Cloud Storage.

Google Cloud Storage

Google Cloud Storage (GCS) is the backbone of data storage within GCP and is actually used behind the scenes for many of the other applications, such as BigQuery. GCS does one thing well: store bytes. Within GCS itself, you can't manipulate that data as you would if your data were in a database.

You can upload up to 5 TB of data per object and have virtually unlimited space within the buckets that are the base structure of GCS. You can then access your objects via its uniform resource identifier (URI) syntax, e.g., `gs://my-bucket/my-object`. The bucket names are globally unique, meaning you don't need to specify the project they sit within. You can also choose to make your objects public with HTTP URLs, meaning GCS could be used to host websites like a web server.

GCS also has fine-grained control over who can read or write objects within buckets. This lets you upload and download in a secure way that is protected by Google's authorization systems. A common way to import data into GCP is to use a service key running in the source system, which will allow it to upload.

GCS is often used as the landing pad for imported data because you don't need to worry about schema or other loading issues that may prevent imports to other systems. As long as the object has bytes, it can be uploaded. For this reason, even if your data is structured and can be loaded into a database directly, it's helpful to have a raw backup in case that loading fails for import schema reasons.

GCS is also the only location to which you can upload unstructured data such as videos, sound files, or images. Many other GCP services assume you have uploaded to GCS as inputs to their services, such as the machine learning APIs that take your GCS URI as an input for operating upon the file, turning it into structured data you can then use later.

You can read more about using GCS in "GCS" on page 122.

Event-Driven Storage

GCS has events that trigger Pub/Sub upon each change or update, such as a new file, deletion, or edit. This event trigger carries the file location and event name, which means receiving systems like Cloud Functions can use that message to trigger actions.

An example of this is loading a CSV file into BigQuery. A third party can be uploading import files that hit GCS every morning, which triggers your import job. This is preferable to a cron schedule since it's resistant to deliveries being later than your schedule time. This allows for differences in export times and is more robust than only running your import script at a set time each morning (no disasters when daylight saving time occurs, for example).

Cloud Storage Pub/Sub triggers into Cloud Functions

Whenever a file is uploaded to Cloud Storage, it creates a `FINALIZE/CREATE` event in Pub/Sub. This can be used to trigger a Cloud Function that will compare the data with the expected data schema for the BigQuery table it will eventually be added to.

The Cloud Function will be Python, which uses the Google APIs to receive the new file's Storage location (its URI) and then uses that filename to start the load job into BigQuery.

For this example, let's set up a Cloud Function that will take a file from a Cloud Bucket called `marks-bucket-of-stuff` and load a BigQuery table in `my-project:my-dataset.my_crm_imports`.

Cloud Storage file structure. When specifying the files that come into Cloud Storage, there are certain requirements that will make your life easier. I usually specify as follows:

- Use UTF-8 encoding only.
- Use an agreed-upon CSV format (e.g., comma- or semicolon-separated fields— usually a bit lenient here as some systems can't support standards).

- Specify whether data is fully quoted and escaped, or fields are not to be quoted.

- Specify lowercase, no spaces, snake_case filenames. As we'll see later, the filenames will govern the names of the BigQuery tables.

- Suffix the files uploaded with the date (YYYYMMDD) or datetime (YYYYMMDDHHSS) as appropriate.

- If data volumes under, say, 10 GB, full data uploads versus incremental.

 Take particular notice of date formats and file encodings from legacy CRM systems. If you can, specify UTF-8, but you may need some creative processing in your Cloud function to deal with other encodings.

This creates filenames like `my_crm_import_20210703.csv`. The date at the end controls which BigQuery date partition it will be written to.

If the data contains what is considered sensitive data (which CRM records typically are), then for data privacy, these files are set to expire using Cloud Storage's lifecycle rules on the bucket (see Figure 3-20).

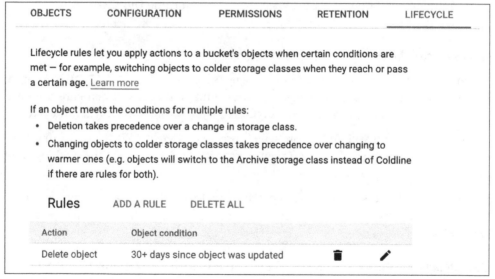

Figure 3-20. GCS lifecycle rules allow you to ensure stray data does not stick around for longer than is welcome

Since GDPR demands responses within 40 days, a 30-day expiration window serves as a buffer for disaster backup but is low enough to respect the desire not to have private data hanging around too long.

Cloud Function example for importing from GCS to BigQuery. Once your files appear in Cloud Storage via the CRM export schedule, you can turn those files into BigQuery tables. You can do this with the Cloud Storage triggers (*https://oreil.ly/0EwhB*).

You can create Cloud Functions via the Cloud Console (*https://oreil.ly/2TBfH*). You will want to select a name and region closest to the location of your bucket and change the trigger type to "Cloud Storage" (see Figure 3-21).

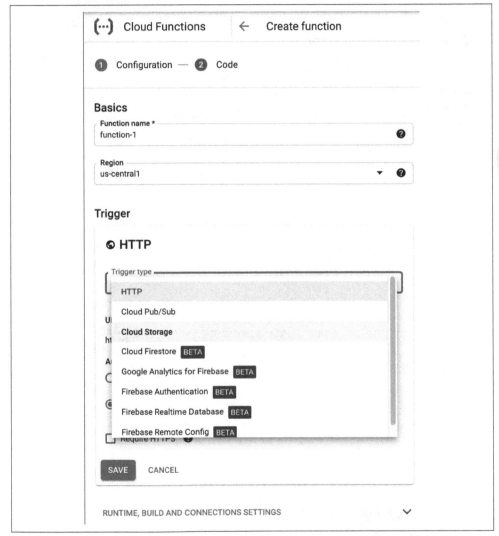

Figure 3-21. Creating a Cloud Function to trigger from a newly uploaded file in GCS

You can then select the Cloud Storage bucket you have access to and the type of trigger. For this use case, we want to know when the file has completed its upload: Finalize/Create (Figure 3-22).

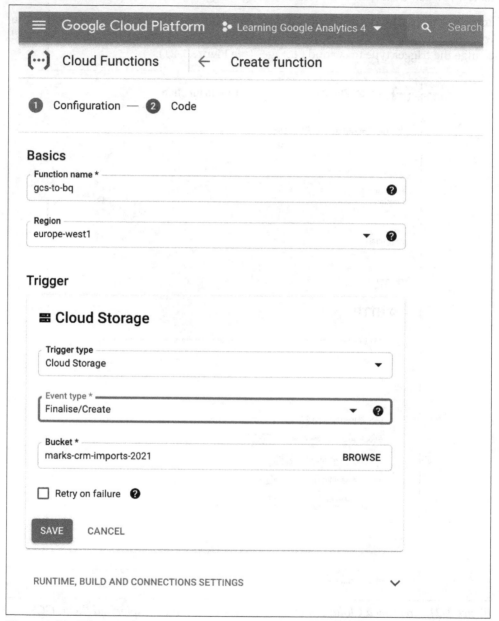

Figure 3-22. Selecting the event type (Finalize/Create) and the bucket that will send the Pub/Sub event

Next we can add the code that will trigger once that event is detected.

Selecting the Runtime to be Python, we'll find some code already filled in to point us in the right direction, as shown in Figure 3-23.

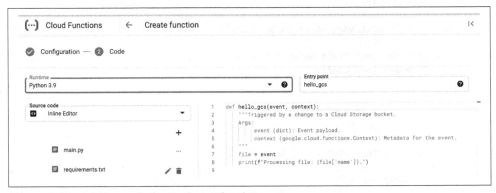

Figure 3-23. Adding code for Python Cloud Functions

The default code in the screenshot will trigger the hello_gcs(event, context) event and print the filename to the logs via event['name']. The event['name'] is all the data we need for the Pub/Sub event to get going with our BigQuery import function. For our own code in Example 3-9, we'll modify it so it will take the name of the file loaded into GCS and parse it into BigQuery. The BigQuery configuration will follow the Python SDK documentation (*https://oreil.ly/9aHTA*).

Example 3-9. Cloud Function Python code to load a CSV from a GCS bucket into BigQuery

```
import os
import yaml
import logging
import re
import datetime
from google.cloud import bigquery
from google.cloud.bigquery import LoadJobConfig
from google.cloud.bigquery import SchemaField
import google.cloud.logging

# set up logging https://cloud.google.com/logging/docs/setup/python
client = google.cloud.logging.Client()
client.get_default_handler()
client.setup_logging()

# load config.yaml into config
config_file = "config.yaml"

if os.path.isfile(config_file):
```

```python
    with open("config.yaml", "r") as stream:
        try:
            config = yaml.safe_load(stream)
        except yaml.YAMLError as exc:
            logging.error(exc)
else:
    logging.error("config.yaml needs to be added")

# create a list of SchemaField objects from a schema config.yaml file
def create_schema(schema_config):

    SCHEMA = []
    for scheme in schema_config:

        if 'description' in scheme:
            description = scheme['description']
        else:
            description = ''

        if 'mode' in scheme:
            mode = scheme['mode']
        else:
            mode = 'NULLABLE'

        try:
            assert isinstance(scheme['name'], str)
            assert isinstance(scheme['type'], str)
            assert isinstance(mode, str)
            assert isinstance(description, str)
        except AssertionError as e:
            logging.info(
                'Error in schema: name {} - type {}
                - mode - {} description {}'.format(scheme['name'], scheme['type'],
                                                    mode, description))

            break

        entry = SchemaField(name=scheme['name'],
                            field_type=scheme['type'],
                            mode=mode,
                            description=description)
        SCHEMA.append(entry)

    logging.debug('SCHEMA created {}'.format(SCHEMA))

    return SCHEMA


def make_tbl_name(table_id, schema=False):

    t_split = table_id.split('_20')
```

```python
    name = t_split[0]

    if schema: return name

    suffix = ''.join(re.findall('\d\d', table_id)[0:4])

    return name + '$' + suffix


def query_schema(table_id, job_config):

    schema_name = make_tbl_name(table_id, schema=True)

    logging.info('Looking for schema_name: {} for import: {}'.format(schema_name,
        table_id))
    # if we have no configuration attempt auto-detection
    # recommended only for development tables
    if schema_name not in config['schema']:
        logging.info('No config found. Using auto detection of schema')
        job_config.autodetect = True
        return job_config

    logging.info('Found schema for ' + schema_name)

    schema_config = config['schema'][schema_name]['fields']

    job_config.schema = create_schema(schema_config)

    # standard csv load behavior can be defined here
    job_config.quote_character = '"'
    job_config.skip_leading_rows = 1
    job_config.field_delimiter = ','
    job_config.allow_quoted_newlines = True

    return job_config

def load_gcs_bq(uri, table_id, project, dataset_id):

    client = bigquery.Client(project=project)
    dataset_ref = client.dataset(dataset_id)

    # Change the below configuration according to your import needs
    job_config = LoadJobConfig()
    job_config.source_format = bigquery.SourceFormat.CSV
    job_config.write_disposition = bigquery.WriteDisposition.WRITE_TRUNCATE
    job_config.encoding = bigquery.Encoding.UTF_8
    job_config.time_partitioning = bigquery.TimePartitioning()

    job_config = query_schema(table_id, job_config)

    table_name = make_tbl_name(table_id)
    table_ref = dataset_ref.table(table_name)
```

```
    job = client.load_table_from_uri(
        uri,
        table_ref,
        location='EU',
        job_config=job_config)  # API request


def gcs_to_bq(data, context):
    """Background Cloud Function to be triggered by Cloud Storage.
       This functions constructs the file URI and uploads it to BigQuery.

    Args:
        data (dict): The Cloud Functions event payload.
        context (google.cloud.functions.Context): Metadata of triggering event.
    Returns:
        None; the output is written to Stackdriver Logging
    """

    object_name = data['name']
    project = config['project']
    dataset_id = config['datasetid']

    if object_name:
        # create a bigquery table related to the filename
        table_id = os.path.splitext(os.path.basename(object_name))[0].replace('.','_')
        uri = 'gs://{}/{}'.format(data['bucket'], object_name)

        load_gcs_bq(uri, table_id, project, dataset_id)

    else:
        logging.info('Nothing to load')

    return
```

The *requirements.txt* file will need to be specified as in Example 3-10, which I've veri-fied is working for the Python 3.9 runtime.

Example 3-10. The requirements.txt *file for which Python modules to load from pip*

```
google-cloud-bigquery==2.20.0
google-cloud-logging==2.5.0
pyyaml==5.4.1
```

 As with all code snippets in this book, I've attempted to make sure this works with the latest versions, but you may have to tweak the code and/or dependency requirements depending on how far in the future you are reading.

The code relies on a *config.yaml* file that you'll look for in the same folder as the Python code. If present, it will use the file to specify the schema for the BigQuery table it creates. If no schema for the filename is found, then it will fall back to autodetection. This allows you to import multiple BigQuery tables. An example of the configuration file is Example 3-11, which uses a YAML format.

Example 3-11. A YAML config file for use with the Cloud Function specified in Example 3-9

```
project: learning-ga4
datasetid: crm_imports
schema:
  crm_bookings:
    fields:
        - name: BOOK_ID
          type: STRING
        - name: BOOKING_ACTIVE
          type: STRING
        - name: BOOKING_DEPOSIT
          type: STRING
        - name: DATE
          type: STRING
        - name: DEPARTURE_DATE
          type: STRING
  crm_permissions:
    fields:
        - name: USER_ID
          type: STRING
        - name: PERMISSION
          type: STRING
        - name: STATUS
          type: STRING
        - name: SOURCE
          type: STRING
        - name: PERMISSION_DATE
          type: STRING
  crm_sales:
    fields:
        - name: SALES_ID
          type: STRING
        - name: SALES_EMAIL
          type: STRING
        - name: SALES_FIRST_NAME
          type: STRING
        - name: SALES_LAST_NAME
          type: STRING
```

The configuration in Example 3-11 shows an example with three tables to import. The advantage of event-based imports is that you can trigger many functions at once.

If a CSV file is uploaded that is not specified in the schema section, then a BigQuery load will be attempted with autodetection of the schema. This is helpful in development, but it's highly recommended to use specific schema in production.

To test it, a CSV can be loaded without (Example 3-12) and with (Example 3-13) the schema you specify.

Example 3-12. An example of a CSV file uploaded that has not been specified in the schema

```
USER_ID,EMAIL,TOTAL_LIFETIME_REVENUE
AB12345,david@email.com,56789
AB34252,sanne@freeemail.com,34234
RF45343,rose@medson.com,23123
```

Example 3-13. An example of a CSV file that matches the configuration schema

```
USER_ID,PERMISSION,STATUS,SOURCE,PERMISSION_DATE
AB12345,Marketing1,True,Email,2021-01-21
AB34252,Marketing3,True,Website,2020-12-02
RF45343,-,False,-,-
```

Create the test CSV files and upload them to the Cloud Storage bucket. Within the logs of the Cloud Function (Figure 3-24), you should see the logs from within the function indicating whether it was triggered and started the BigQuery job.

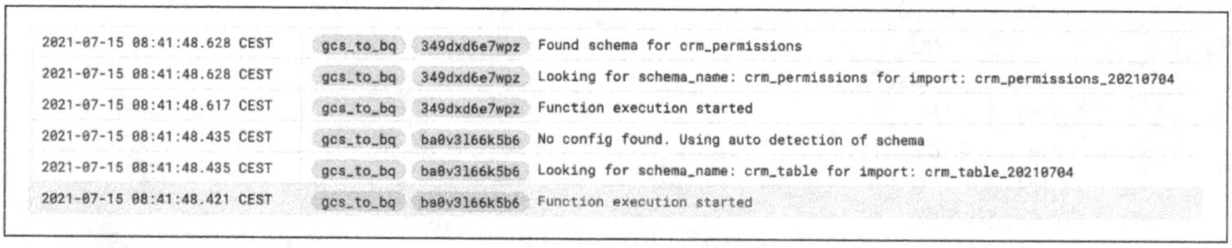

2021-07-15 08:41:48.628 CEST	gcs_to_bq	349dxd6e7wpz	Found schema for crm_permissions
2021-07-15 08:41:48.628 CEST	gcs_to_bq	349dxd6e7wpz	Looking for schema_name: crm_permissions for import: crm_permissions_20210704
2021-07-15 08:41:48.617 CEST	gcs_to_bq	349dxd6e7wpz	Function execution started
2021-07-15 08:41:48.435 CEST	gcs_to_bq	ba0v3166k5b6	No config found. Using auto detection of schema
2021-07-15 08:41:48.435 CEST	gcs_to_bq	ba0v3166k5b6	Looking for schema_name: crm_table for import: crm_table_20210704
2021-07-15 08:41:48.421 CEST	gcs_to_bq	ba0v3166k5b6	Function execution started

Figure 3-24. Inspecting the Cloud Function logs we can see one file was imported to a specified schema and one used autodetection

This is only half the story, though—you should also check the BigQuery logs to see whether the BigQuery load job was successful. In Figure 3-25, we see that the schema was included and a successful BigQuery load job operated.

```
  ▼ job: {
      ▼ jobConfiguration: {
        ▼ load: {
            createDisposition: "CREATE_IF_NEEDED"
          ▼ destinationTable: {
              datasetId: "crm_imports"
              projectId: "learning-ga4"
              tableId: "crm_permissions$20210704"
            }
            schemaJson: "{
                        "fields": [{
                          "name": "USER_ID",
                          "type": "STRING",
                          "mode": "NULLABLE"
                        }, {
                          "name": "PERMISSION",
                          "type": "STRING",
                          "mode": "NULLABLE"
                        }, {
                          "name": "STATUS",
                          "type": "STRING",
                          "mode": "NULLABLE"
                        }, {
                          "name": "SOURCE",
                          "type": "STRING",
                          "mode": "NULLABLE"
                        }, {
                          "name": "PERMISSION_DATE",
                          "type": "STRING",
                          "mode": "NULLABLE"
                        }]
                      }"
          ▼ sourceUris: [
              0: "gs://marks-crm-imports-2021/crm_permissions_20210704.csv"
            ]
```

*Figure 3-25. Inspecting the BigQuery logs to see the schema has been specified as
expected*

After checking all the logs, you should see the tables within BigQuery itself with the
schema as specified (or not, if using autodetection). See Figure 3-26.

To use the script we have developed within your own pipelines, you now need to create your CSV exports and create the *config.yaml* file to suit your own data. You can
also deploy multiple cloud functions with different configuration settings suited to
each CRM export or destination. The main goal is achieved: we now have event-
driven imports into BigQuery for CSVs that could be up to 5 TB, or 4 GB if you
directly load gzipped files. See the documentation (*https://oreil.ly/sGo2O*) for more
details.

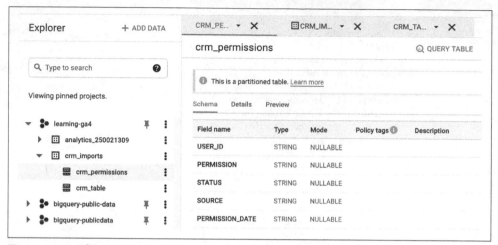

Figure 3-26. The BigQuery tables are imported from the CSV on GCS with the specified schema

Data Privacy

From a data privacy perspective, in GCS you can set expiration times for data that will allow you to safely delete any personal data. You can use this with regular imports to set an expiration that is under the legal response times of data requests. This means the source system's existing data deletion procedure can be maintained without you needing to replicate it across the cloud; a user who asks that their data be deleted in the existing system will have that request filter through to the cloud data within, say, 30 days.

Privacy comes up much more often when you're dealing with personal data that is often within internal databases such as CRM, which we'll talk about in the next section.

CRM Database Imports via GCS

Because I can't become an expert on every database out there, I usually let clients know that I'll be responsible for their data exports once they hit Cloud Storage, but they have to be responsible for getting the exports there in the first place. This is usually fine since the actual requests to the development team are of a simple nature, like exporting columns A, B, and C into a CSV or JSON file and scheduling uploads to GCS using gcloud or one of the Cloud Storage SDKs (*https://oreil.ly/Km9XH*). If you are in-house, you may be more involved in how the actual exports are created and delivered from something like a local MySQL database.

Specifying the upload to Cloud Storage and not directly into BigQuery makes uploading easy since the export team doesn't need to comply with any specific schema. That will be your job when you load from Cloud Storage. This also provides a handy raw data backup.

The export script from the local CRM database is best done using a service-key authentication file that has been restricted to Cloud Storage bucket roles only. Once the data hits Cloud Storage, you can use a Cloud Function as outlined in "Event-Driven Storage" on page 83 to load the data into BigQuery.

If you've implemented all of the code and functions in this chapter so far, you may be experiencing quite a lot of overhead copy-pasting code in and out of Cloud Functions and other sources, which is more familiar when working with GTM or similar. However, this method is prone to errors and can waste your time looking for historic changes. A much more developer-friendly way to deploy code is by following software engineering best practices such as CI/CD, which is covered in the next section about Google Cloud's service: Cloud Build.

Setting Up Cloud Build CI/CD with GitHub

I'll include the Cloud Build Git triggers here as they will be helpful for some data ingestion tasks, but truth be told, they will be helpful throughout the data pipeline. To get a better outline, see "Cloud Build" on page 137 in the data storage section, which goes into more detail.

It's a good idea to set Cloud Build up as early as possible in the process because it speeds up development and makes it much easier in the long run.

Setting Up GitHub

I'll use GitHub as an example, but any Git system is supported by mirroring with Google's own Git system, Source Repositories. Instructions for using Git/GitHub are beyond the scope of this book, but GitHub offers some resources that can help (*https://oreil.ly/wm0qC*).

Once you have GitHub, create an empty repository that will hold all the files for the project. You then need to activate the GitHub app for Cloud Build (*https://oreil.ly/C4mue*). You can choose to let it access all (so you don't need to configure again in the future) or selected repositories. Make sure your choice covers the repository you just created.

Setting Up the GitHub Connection to Cloud Build

We need to give permission for Cloud Build to trigger events from GitHub. Within the Google Cloud Console, go to the Cloud Build section (*https://oreil.ly/CFddT*) and link to the repository. It will give you the option to add any that have the GitHub app activated. See Figure 3-27.

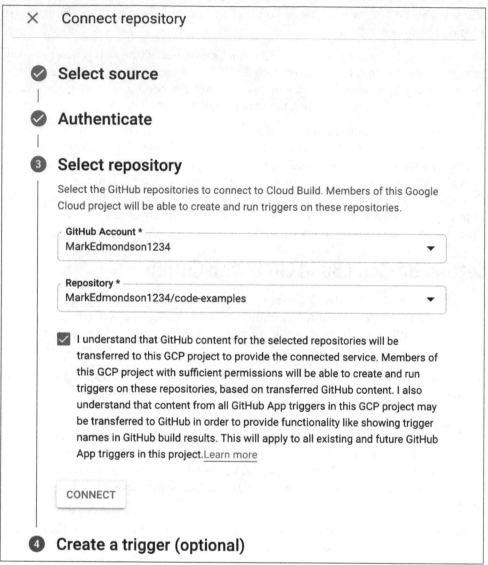

Figure 3-27. Linking Cloud Build to the GitHub repository holding your files

You can also now choose to make a trigger, which will govern the rules about activating the Cloud Build steps created in *cloudbuild.yaml* detailed in "Adding Files to the Repository" on page 98.

After creating a trigger, we want it to deploy only when the files related to the Cloud Function are changed. An example is shown in Figure 3-28.

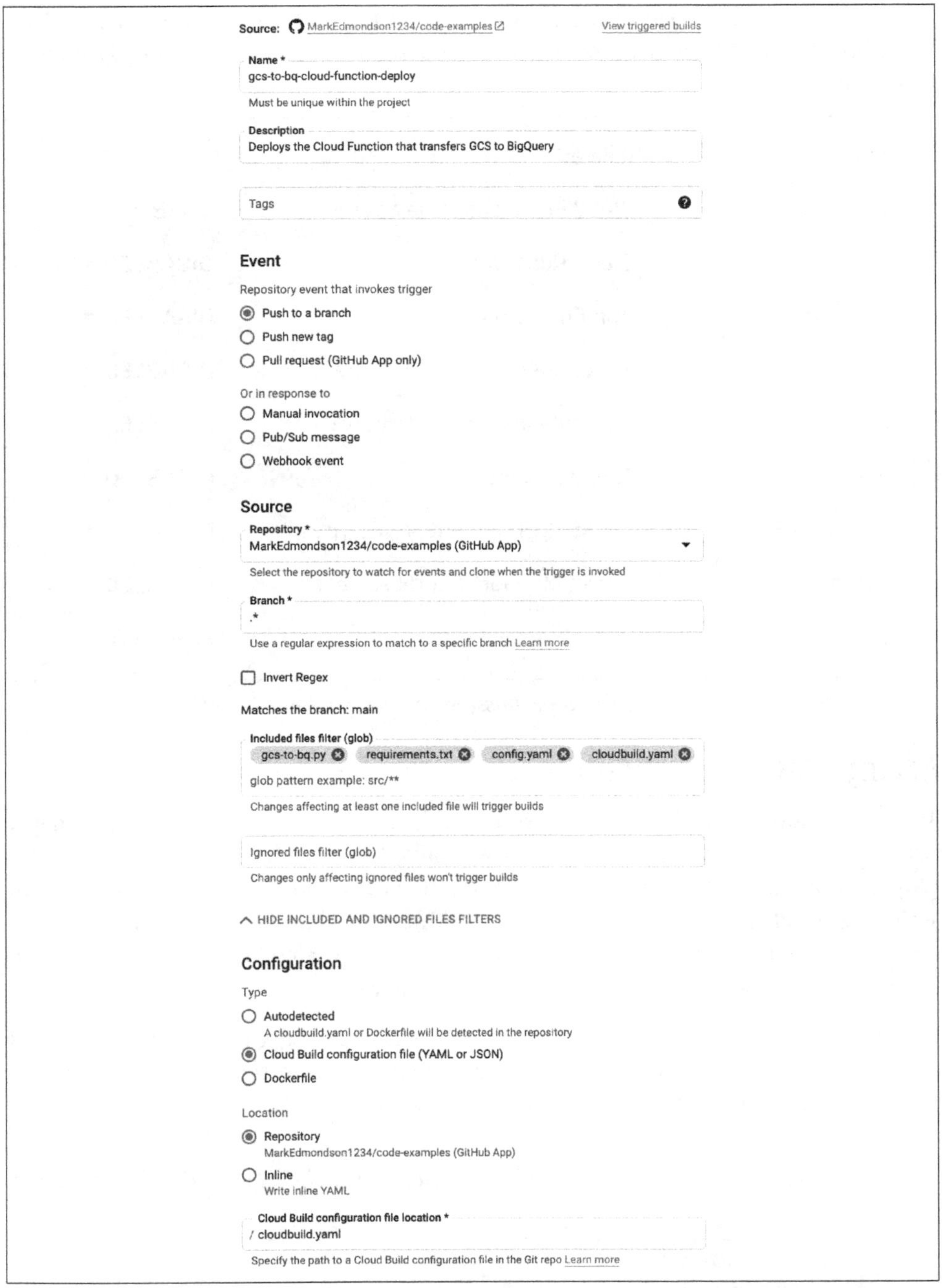

Figure 3-28. A Cloud Build trigger that will activate the contents of the cloudbuild.yaml *upon each commit to the GitHub repository*

In this case, we'll also need to add the Cloud Build builder agent as a user authorized to deploy Cloud functions, which can be set in the settings pane (*https://oreil.ly/6JeRG*) shown in Figure 3-29.

GCP service	Role ?	Status
Cloud Functions	Cloud Functions Developer	ENABLED ▾
Cloud Run	Cloud Run Admin	DISABLED ▾
App Engine	App Engine Admin	DISABLED ▾
Kubernetes Engine	Kubernetes Engine Developer	DISABLED ▾
Compute Engine	Compute Instance Admin (v1)	DISABLED ▾
Firebase	Firebase Admin	DISABLED ▾
Cloud KMS	Cloud KMS CryptoKey Decrypter	DISABLED ▾
Secret Manager	Secret Manager Secret Accessor	DISABLED ▾
Service Accounts	Service Account User	ENABLED ▾

Figure 3-29. Setting the Cloud Build permission to deploy Cloud Functions

Adding Files to the Repository

To enable Cloud Build, we need an extra file, *cloudbuild.yaml*, which governs what it will build. In this example, any changes we make to the code should trigger a redeployment of the Cloud Function. Consulting the Cloud Function documentation, we see that `gcloud`, a command-line tool for working with GCP, is recommended to trigger deployments when not using the web console. In particular, we need the `gcloud functions deploy` (*https://oreil.ly/0pOzR*) commands.

For our Cloud Function and triggers, this translates to running the gcloud commands shown in Example 3-14. The job of our *cloudbuild.yaml* is to replicate this command to trigger each time the Cloud Function code changes.

Example 3-14. `gcloud` command to deploy the Cloud Function specified in Example 3-9

```
gcloud functions deploy gcs_to_bq \
    --runtime=python39 \
    --region=europe-west1 \
    --trigger-resource=marks-crm-imports-2021 \
    --trigger-event=google.storage.object.finalize
```

Translating the deployment code to Cloud Build YAML format, the YAML shown in Example 3-15 is produced.

Example 3-15. Cloud Build YAML for deploying a Cloud Function from Example 3-9

```
steps:
- name: gcr.io/cloud-builders/gcloud
  args: ['functions',
         'deploy',
         'gcs_to_bq',
         '--runtime=python39',
         '--region=europe-west1',
         '--trigger-resource=marks-crm-imports-2021',
         '--trigger-event=google.storage.object.finalize']
```

Example 3-15 uses a docker image in `name: gcr.io/cloud-builders/gcloud` that has `gcloud` installed (as you can imagine, this is a handy Docker image to have around). The command will by default be run in the root folder of the GitHub repository we've assigned to it, so it will deploy all the files present, including any configuration or *requirements.txt* dependency files.

All the files from "Event-Driven Storage" on page 83 plus the *cloudbuild.yaml* file should now be included in the Git repository, as shown in Figure 3-30.

Ensure that your file containing the Cloud Function is called *main.py*.

Name		Date Modified	Size	Kind
cloudbuild.yaml		Today at 13.39	329 bytes	YAML
config.yaml		Today at 12.33	844 bytes	YAML
crm_permissions_20210708.csv		Today at 12.34	152 bytes	CSV Document
crm_permissions.csv		Today at 12.35	152 bytes	CSV Document
crm_users_20210708.csv		Today at 12.33	130 bytes	CSV Document
main.py		Today at 12.32	4 KB	Python
README.md		Today at 12.39	250 bytes	R Markdown File
requirements.txt		Today at 12.33	72 bytes	Plain Text

Figure 3-30. Files in a folder enabled for Git

Committing the files and pushing up to GitHub should then trigger a build if the trigger was set up correctly in "Setting Up the GitHub Connection to Cloud Build" on page 95. You can then view the progress of your build (*https://oreil.ly/5xl9E*), which is useful to see if the syntax is correct. A successful build will appear with a nice tick mark (Figure 3-31). You should confirm that the Cloud Function is then deployed with your changes.

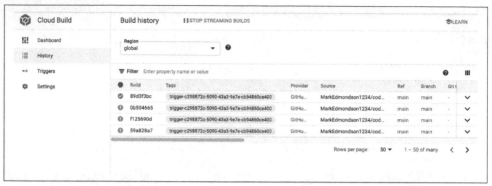

Figure 3-31. A successful deployment of a Cloud Function via Cloud Build; note the failed builds from before—we all make mistakes!

From now on, you should never need use the WebUI to make changes to your Cloud Function. Make changes locally to your development files and commit them to GitHub, and in a few minutes those changes should be reflected in your deployed code. This offers vast speedups and enables you to iterate much more quickly.

Summary

This chapter has gone over the data ingestions you will need for all the use cases in the book, and probably for around 95% of real-world applications. Once you have your GA4 data streams set up, the combination of BigQuery, Cloud Storage, GTM, and internal systems opens up a wealth of rich datasets.

If you're looking for extra skills to help enhance your GA4 setup, becoming familiar with the Cloud Platform services as a great next step since you will then have the tools to really expand GA4's horizons. This was a key step in my career in delivering some exciting projects.

There is a lot more we could cover, but I hope the tour in this chapter has introduced the common ways to get data into your systems. The next chapter, Chapter 4, will show you how to work with that data once it's there, which is a necessary next step before moving to the data modeling and data activation stages. We've touched on some of these elements already with Cloud Storage and BigQuery, but the next chapter will get into more principles and introduce other systems that may complement those, in particular for real-time or scheduled streams.

## CHAPTER 4

Data Storage

Where you store the data for your application is a critical part of your data analytics infrastructure. It can vary from a trivial concern, where you simply use GA4's native storage systems, to complex data flows where you are ingesting multiple data sources including GA4, your CRM database, other digital marketing channel cost data, and more. Here, BigQuery as the analytics database of choice in GCP really dominates because it has been built to tackle exactly the type of issues that come up when considering working with data from an analytics perspective, which is exactly why GA4 offers it as an option to export. In general, the philosophy is to bring all your data into one location where you can run analytics queries over it with ease and make it available to whichever people or applications need it in a security-conscious but democratic way.

This chapter will go over the various decisions and strategies I have learned to consider when dealing with data storage systems. I want you to benefit from my mistakes so you can avoid them and set yourself up with a solid foundation for any of your use cases.

This chapter is the glue between the data collection and data modeling parts of your data analytics projects. Your GA4 data should be flowing in under the principles laid out in Chapter 3, and you will work with it with the tools and techniques described in this chapter with the intention of using the methods described in Chapters 5 and 6, all guided by the use cases that Chapter 2 helped you define.

We'll start with some of the general principles you should consider when looking at your data storage solution, and then run through some of the most popular options on GCP and those I use every day.

Data Principles

This section will go over some general guidelines to guide you in whatever data storage options you're using. We talk about how to tidy and keep your data at a high standard, how to fashion your datasets to suit different roles your business may need, and things to think about when linking datasets.

Tidy Data

Tidy data is a concept introduced to me from within the R community and is such a good idea that all data practitioners can benefit from following its principles. Tidy data is an opinionated description of how you should store your data so that it is the most useful for downstream data flows. It looks to give you set parameters for how you should be storing your data so that you have a common foundation for all of your data projects.

The concept of tidy data was developed by Hadley Wickham, Data Scientist at RStudio and inventor of the concept for the "tidyverse." See the *R for Data Science: Import, Tidy, Transform, Visualize, and Model Data* by Garrett Grolemund and Hadley Wickham (O'Reilly) for a good grounding on applying its principles, or visit the tidyverse website (*https://www.tidyverse.org*).

Although tidy data first became popular within R's data science community, even if you don't use R I recommend thinking about it as a first goal in your data processing. This is encapsulated in a quote from Wickham and Grolemund's book (*https:// oreil.ly/Z4Faj*):

> Happy families are all alike; every unhappy family is unhappy in its own way.
>
> —Leo Tolstoy

> Tidy datasets are all alike, but every messy dataset is messy in its own way.
>
> —Hadley Wickham

The fundamental idea is that there is a way to turn your raw data into a universal standard that will be useful for data analysis down the line, and if you apply this to your data, you won't need to reinvent the wheel each time you want to process your data.

There are three rules that, if followed, will make your dataset tidy:

1. Each variable must have its own column.

2. Each observation must have its own row.

3. Each value must have its own cell.

You can see these rules illustrated in Figure 4-1.

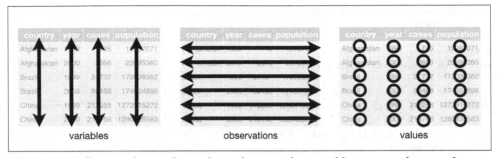

Figure 4-1. Following three rules makes a dataset tidy: variables are in columns, observations are in rows, and values are in cells (from R for Data Science *by Wickham and Grolemund)*

Because data cleaning is usually the most time-consuming part of a project, this frees up a lot of mental capacity to work on the specific problems for your use case without having to worry about the shape your data is in every time you start. I recommend that after you import your raw data, you should make every effort to create tidy data versions of that data that are the ones you expose to downstream use cases. The tidy data standard helps you by taking out the brainwork of thinking about how your data should be shaped each time and lets your downstream data applications standardize since they can expect data will always come in a particular manner.

That's the theory, so let's see how this works in practice in the next section.

Example of tidying GA4 data

An example follows for a workflow where we start with untidy GA4 data and clean it up so that it is ready for further analysis. We're using R for the tidying, but the same principles can apply for any language or tool such as Excel.

Let's start with some GA4 data. Example 4-1 shows an R script that will export some customEvent data from my blog. This data includes the category I have put each blog post in, such as "Google Analytics" or "BigQuery." This custom data is available in a customEvent named `category`.

Example 4-1. R script for extracting the custom dimension `category` *from the GA4 Data API*

```
library(googleAnalyticsR)

# authenticate with a user with access
ga_auth()

# if you have forgotten your propertyID
ga4s <- ga_account_list("ga4")
```

```
# my blog propertyId - change for your own
gaid <- 206670707

# import your custom fields
meta <- ga_meta("data", propertyId = gaid)

# date range of when the field was implemented to today
date_range <- c("2021-07-01", as.character(Sys.Date()))

# filter out any data that doesn't have a category
invalid_category <-
 ga_data_filter(!"customEvent:category" == c("(not set)","null"))

# API call to see trend of custom field: article_read
article_reads <- ga_data(gaid,
    metrics = "eventCount",
    date_range = date_range,
    dimensions = c("date", "customEvent:category"),
    orderBys = ga_data_order(+date),
    dim_filters = invalid_category,
    limit = -1)
```

The top of the contents of `article_reads` is shown in Table 4-1.

You can see that the quality of data collection has a knock-on effect for downstream data processing: for instance, the article category could have been split out into its own events to make the data cleaner. It is not "tidy" data. We'll need to clean the data to make it suitable for modeling—this is extremely common. It also highlights how clean data capture can reduce work downstream.

Table 4-1. GA4 data extracted via the Data API via googleAnalyticsR

date	customEvent:category	eventCount
2021-07-01	GOOGLE-TAG-MANAGER · CLOUD-FUNCTIONS	13
2021-07-01	GOOGLE-TAG-MANAGER · GOOGLE-ANALYTICS	12
2021-07-01	R · GOOGLE-APP-ENGINE · DOCKER · GOOGLE-ANALYTICS · GOOGLE-COMPUTE-ENGINE · RSTUDIO-SERVER	9
2021-07-01	R · CLOUD-RUN · GOOGLE-TAG-MANAGER · BIG-QUERY	8
2021-07-01	R · DOCKER · CLOUD-RUN	8
2021-07-01	GOOGLE-TAG-MANAGER · DOCKER · CLOUD-RUN	7
2021-07-01	R · GOOGLE-ANALYTICS · SEARCH-CONSOLE	7
2021-07-01	R · DOCKER · RSTUDIO-SERVER · GOOGLE-COMPUTE-ENGINE	6
2021-07-01	DOCKER · R	5
2021-07-01	R · FIREBASE · GOOGLE-AUTH · CLOUD-FUNCTIONS · PYTHON	5
2021-07-01	R · GOOGLE-AUTH · BIG-QUERY · GOOGLE-ANALYTICS · GOOGLE-CLOUD-STORAGE · GOOGLE-COMPUTE-ENGINE · GOOG	4
2021-07-01	GOOGLE-CLOUD-STORAGE · PYTHON · GOOGLE-ANALYTICS · CLOUD-FUNCTIONS	3

date	customEvent:category	eventCount
2021-07-01	R · GOOGLE-ANALYTICS	3
2021-07-01	BIG-QUERY · PYTHON · GOOGLE-ANALYTICS · CLOUD-FUNCTIONS	2
2021-07-01	DOCKER · R · GOOGLE-COMPUTE-ENGINE · CLOUD-RUN	2
2021-07-01	R · GOOGLE-AUTH	2
2021-07-01	docker · R	2
2021-07-02	R · CLOUD-RUN · GOOGLE-TAG-MANAGER · BIG-QUERY	9
2021-07-02	DOCKER · R	8
2021-07-02	GOOGLE-TAG-MANAGER · DOCKER · CLOUD-RUN	8
2021-07-02	GOOGLE-TAG-MANAGER · GOOGLE-ANALYTICS	8
2021-07-02	R · DOCKER · CLOUD-RUN	6
2021-07-02	R · GOOGLE-APP-ENGINE · DOCKER · GOOGLE-ANALYTICS · GOOGLE-COMPUTE-ENGINE · RSTUDIO-SERVER	6

As detailed in "Tidy Data" on page 102, this data is not yet in a tidy form ready for analysis, so we'll leverage some of R's `tidyverse` libraries to help clean it up, namely `tidyr` and `dplyr`.

The first job is to rename the column names and separate out the category strings so we have one per column. We also make everything lowercase. See Example 4-2 for how to do this using the tidyverse, given the `article_reads` data.frame from Table 4-1.

Example 4-2. Tidying up the article_reads raw data using `tidy` and `dplyr` so that it looks similar to Table 4-2

```
library(tidyr)
library(dplyr)

clean_cats <- article_reads |>
    # rename data columns
    rename(category = "customEvent:category",
        reads = "eventCount") |>
    # lowercase all category values
    mutate(category = tolower(category)) |>
    # separate the single category column into six
    separate(category,
        into = paste0("category_",1:6),
        sep = "[^[:alnum:]-]+",
        fill = "right", extra = "drop")
```

The data now looks like Table 4-2. However, we're not quite there yet in the tidy format.

Table 4-2. The result of data tidying Table 4-1

date	category_1	category_2	category_3	category_4	category_5	category_6	reads
2021-07-01	google-tag-manager	cloud-functions	NA	NA	NA	NA	13
2021-07-01	google-tag-manager	google-analytics	NA	NA	NA	NA	12
2021-07-01	r	google-app-engine	docker	google-analytics	google-compute-engine	rstudio-server	9
2021-07-01	r	cloud-run	google-tag-manager	big-query	NA	NA	8
2021-07-01	r	docker	cloud-run	NA	NA	NA	8
2021-07-01	google-tag-manager	docker	cloud-run	NA	NA	NA	7

We would like to aggregate the data so each row is a single observation: the number of reads per category per day. We do this by pivoting the data into a "long" format versus the "wide" format we have now. Once the data is in that longer format, aggregation is done over the date and category columns (much like SQL's GROUP BY) to get the sum of reads per category. See Example 4-3.

Example 4-3. Turning the wide data into long and aggregation per date/category

```
library(dplyr)
library(tidyr)

agg_cats <- clean_cats |>
    # turn wide data into long
    pivot_longer(
        cols = starts_with("category_"),
        values_to = "categories",
        values_drop_na = TRUE
    ) |>
    # group over dimensions we wish to aggregate over
    group_by(date, categories) |>
    # create a category_reads metric: the sum of reads
    summarize(category_reads = sum(reads), .groups = "drop_last") |>
    # order by date and reads, descending
    arrange(date, desc(category_reads))
```

> The R examples in this book assume R v4.1, which includes the pipe operator |>. In R versions before 4.1, you will see the pipe operator imported from its own package, magrittr, and it will look like %>%. They can be safely interchanged for these examples.

Once the data is run through the tidying, we should see a table similar to Table 4-3. This is a tidy dataset that any data scientist or analyst should be happy to work with and the starting point of the model exploration phase.

Table 4-3. Tidied data from the `article_reads` raw data

date	categories	category_reads
2021-07-01	r	66
2021-07-01	google-tag-manager	42
2021-07-01	docker	41
2021-07-01	google-analytics	41
2021-07-01	cloud-run	25
2021-07-01	cloud-functions	23

The examples given in books always seem to be idealized, and I feel that they rarely reflect the work you'll need to do on a regular basis, likely many iterations of experimentation, bug fixing, and regex. Although a minified example, it still took me a few attempts to get exactly what I was looking for in the preceding examples. However, this is easier to handle if you keep the tidy data principle in mind—it gives you something to aim for that will likely prevent you from having to redo it later.

My first step after I've collected raw data is to see how to shape it into tidy data like that in our example. But even if the data is tidy, you'll also need to consider the role of that data, which is what we cover in the next section.

Datasets for Different Roles

The raw data coming in is rarely in a state that should be used for production or even exposed to internal end users. As the number of users increases, there will be more reason to prepare tidy datasets for those purposes, but you should keep a "source of truth" so that you can always backtrack to see how the more derived datasets were created.

Here you may need to start thinking about your data governance, which is the process of looking to determine who and what is accessing different types of data.

A few different roles are suggested here:

Raw data

It's a good idea to keep your raw data streams together and untouched so you always have the option to rebuild if anything goes wrong downstream. In GA4's case, this will be the BigQuery data export. It's generally not advised to modify this data via additions or subtractions unless you have legal obligations such as personal data deletion requests. It's also not recommended to expose this dataset to end users unless they have a need since the raw datasets are typically quite hard to work with. For example, the GA4 export is in a nested structure that has a tough learning curve for anyone not yet experienced in BigQuery SQL. This is unfortunate because for some people it's their first taste of data engineering, and they come away thinking it's a lot harder than it is than if, say, they were working with only the more typical flat datasets. Instead, your first workflows will generally take this raw data and tidy, filter, and aggregate it into something a lot more manageable.

Tidy data

This is data that has gone through a first pass of making it fit for consumption. Here you can take out bad data points, standardize naming conventions, perform dataset joins if helpful, produce aggregation tables, and make the data easier to use. When you're looking for a good dataset to serve as the "source of truth," then the tidy data datasets are preferable to the original raw data source. Maintaining this dataset is an ongoing task, most likely done by the data engineers who created it. Downstream data users should have only read access and may help by suggesting useful tables to be included.

Business cases

Included in the many aggregations you can build from the tidy data are the typical business use cases that will be the source for many of your downstream applications. An example would be a merge of your cost data from your media channels and your GA4 web stream data, combined with your conversion data in your CRM. This is a common desired dataset that has the full "closed loop" of marketing effectiveness data within it (cost, action, and conversion). Other business cases may be more focused on sales or product development. If you have enough data, you could make datasets available to the suitable departments on an as-needed basis, which will then be the data source for the day-to-day ad hoc queries an end user may have. The end user will probably access their data either with some limited SQL knowledge or via a data visualization tool such as Looker, Data Studio, or Tableau. Having these relevant datasets available to all in your company is a good signal that you actually are a "data driven organization" (a phrase I think around 90% of all CEOs aspire to but maybe only 10% actually realize).

Test playground

You'll also often need a scratch pad to try out new integrations, joins, and development. Having a dedicated dataset with a data expiration date of 90 days, for example, means you can be assured that people can work within your datasets without you needing to chase down stray test data on users or damaging production systems.

Data applications

Each data application you have running in production is most likely a derivation of all of the previously mentioned dataset roles. Making sure you have a dedicated dataset to your business critical use cases means you can always know exactly what data is being used and avoid other use cases interfering with yours down the line.

These roles are in a rough order of data flows. It's typical that views or scheduled tasks are set up to process and copy data over to their respective dependents, and you may have them in different GCP projects for administration.

 A big value when using datasets such as GA4's BigQuery exports will be linking that data to your other data, discussed in "Linking Datasets" on page 175.

We have explored some of what will help make your datasets a joy to work with for your users. Should you realize the dream of tidy, role-defined datasets that link data across your business departments in a way that users have all they need at a touch of a button (or SQL query), then you will already be ahead of a vast number of businesses. As an example, consider Google, which many would consider the epitome of a data driven company. In Lak's book, *Data Science on the Google Cloud Platform*, he recounts how 80% of Google employees use data on a weekly basis:

> At Google, for example, nearly 80% of employees use Dremel (Dremel is the internal counterpart to Google Cloud's BigQuery) each month. Some use data in more sophisticated ways than others, but everyone touches data on a regular basis to inform their decisions. Ask someone a question, and you are likely to receive a link to a BigQuery view of query rather than to the actual answer: "Run this query every time you want to know the most up-to-date answer," goes the thinking. BigQuery in the latter scenario has gone from being the no-ops database replacement to being the self-serve data analytics solution.

The quote reflects what a lot of companies work toward and wish to be available to their own employees, and this would have a large business impact if fully realized.

In the next section, we'll consider the tool mentioned in the quote that enabled it for Google: BigQuery.

BigQuery

It's somewhat of a truism that your data analytics needs will all be solved if you just use BigQuery. It certainly has had a big impact on my career and turned data engineering from a frustrating exercise of spending a large amount of time on infrastructure and loading tasks into being able to concentrate more time on getting value out of data.

We've already talked about BigQuery in "BigQuery" on page 76 in the data ingestion section, with regard to the GA4 BigQuery exports ("Linking GA4 with BigQuery" on page 76) and importing Cloud Storage files from Cloud Storage for use with CRM exports ("Event-Driven Storage" on page 83). This section discusses how to organize and work with your data now that it's sitting in BigQuery.

When to Use BigQuery

It's perhaps easier to outline when not to use BigQuery, since it is somewhat of a panacea for digital analytics tasks on GCP. BigQuery has the following features, which you also want for an analytics database:

- Cheap or free storage so you can throw in all your data without worrying about costs.

- Infinite scale so you don't have to worry about creating new instances of servers to bind together later when you throw in even petabytes worth of data.

- Flexible cost structures: the usual choice is one that scales up only as you use it more (via queries) rather than a sunk cost each month paying for servers, or you can choose to reserve slots for a sunk cost for query cost savings.

- Integrations with the rest of your GCP suite to enhance your data via machine learning or otherwise.

- In-database calculations covering common SQL functions such as COUNT, MEANS, and SUMs, all the way up to machine learning tasks such as clustering and forecasting, meaning you don't need to export, model, then put data back in.

- Massively scalable window functions that would make a traditional database crash.

- Quick returns on your results (minutes verses hours in traditional databases) even when scanning through billions of rows.

- A flexible data structure that lets you work with many-to-one and one-to-many data points without needing many separate tables (the data nesting feature).

- Easy access via a web interface with a secure OAuth2 login.

- Fine-grained user access features from project, dataset, and table down to the ability to give user access only to individual rows and columns.

- A powerful external API covering all features that allows you to both create your own applications and to choose third-party software that has used the same API to create helpful middleware.

- Integration with other clouds such as AWS and Azure to import/export your existing data stacks—for example, with BigQuery Omni you can query data directly on other cloud providers.

- Streaming data applications for near real-time updates.

- Ability to auto-detect data schema and be somewhat flexible when adding new fields.

BigQuery has these features because it's been designed to be the ultimate analytics database, whereas more traditional SQL databases focused on quick row transactional access that sacrifices speed when looking over columns.

BigQuery was one of the first cloud database systems dedicated to analytics, but as of 2022, there are several other database platforms that offer similar performance, such as Snowflake, which is making the sector more competitive. This is driving innovation in BigQuery and beyond, and this can only be a good thing for users of any platform. Regardless, the same principles should apply. Before getting into the nuts and bolts of the SQL queries, we'll now turn to how datasets are organized within BigQuery.

Dataset Organization

I've picked up a few principles from working with BigQuery datasets that may be useful to pass on here.

The first consideration is to locate your dataset in a region that is relevant for your users. One of the few restrictions of BigQuery SQL is that you cannot join data tables across regions, which means that your EU- and US-based data will not be easily merged. For example, if working from the EU, this usually means you need to specify the EU region when creating datasets.

 By default, BigQuery assumes you want your data in the US. It's recommended that you always specify the region when creating your dataset just so you can be sure where it sits and so you won't have to perform a region transfer for all your data later. This is particularly relevant for privacy compliance issues.

A good naming structure for your datasets is also useful so users can quickly find the data they're looking for. Examples include always specifying the source and role of that dataset, rather than just numeric IDs: `ga4_tidy` rather than the GA4 MeasurementId `G-1234567`.

Also, don't be afraid to put data in other GCP projects if it makes sense organizationally—BigQuery SQL works across projects, so a user who has access to both projects will be able to query them (if both tables are in the same region). A common application of this is to have dev, staging, and production projects. A suggested categorization of your BigQuery datasets follow, the main themes of this book:

Raw datasets
Datasets that are the first destination for external APIs or services.

Tidy datasets
Datasets that are tidied up and perhaps have aggregations or joins performed to get to a base useful state that other derived tables will use as the "source of truth."

Modeling datasets
Datasets covering the model results that will usually have the tidy datasets as a source and may be intermediate tables for the activation tables later on.

Activation datasets
Datasets that carry the Views and clean tables created for any activation work such as dashboards, API endpoints, or external provider exports.

Test/dev datasets
I usually create a dataset with a data expiration time set to 90 days for development work, giving users a scratch pad to make tables without cluttering up the more production-ready datasets.

With a good dataset naming structure, you're taking the opportunity to add useful metadata to your BigQuery tables that will let the rest of your organization find what they're looking for quickly and easily, reduce training costs, and allow more self-management of your data analysts.

So far we've covered dataset organization, but we now turn to the technical specifications of the tables within those datasets.

Table Tips

This section covers some lessons I've learned when working with tables within BigQuery. It covers strategies to make the job of loading, querying, and extracting data easier. Following these tips when working with your data will set you up for the future:

Partition and cluster when possible
If you're dealing with regular data updates, it's preferable to use partitioned tables, which separate your data into daily (or hourly, monthly, yearly, etc.) tables. You'll then be able to query across all of your data easily but still have performance to limit tables to certain time ranges when needed. Clustering is another related feature of BigQuery that allows you to organize the data so that you can query it faster—you can set this up upon import of your data. You can read more

about both and how they affect your data in Google's "Introduction to Partitioned Tables" (*https://oreil.ly/0zcUK*).

Truncate not append

When importing data, I try to avoid the APPEND model of adding data to the dataset, favoring a more stateless WRITE_TRUNCATE (e.g., overwrite) strategy. This allows reruns without needing to delete any data first, e.g., an idempotent workflow that is stateless. This works best with sharded or partitioned tables. This may not be possible if you're importing very large amounts of data and it's too costly to create a full reload.

Flat as a default but nested for performance

When giving tables to less-experienced SQL users, a flat table is a lot easier for them to work with than the nested structure BigQuery allows. A flat table may be a lot larger than a raw nested table, but you should be aggregating and filtering anyway to help lower the data volume. Nested tables, however, are a good way to ensure you don't have too many joins across data. A good rule of thumb is that if you're always joining your dataset with another, then perhaps that data would be better shaped in a nested structure. These nested tables are more common in the raw datasets.

Implementing these tips means that when you need to rerun an import, you won't need to worry about duplicating data. The incorrect day will be wiped over and the new fresh data will be in its place, but only for that partition so you can avoid having to reimport your whole dataset to be sure of your source of truth.

Costs of SELECT *

I would go so far as to have a rule of thumb to never use SELECT* in your production tables, since it can quickly rack up a lot of costs. This is even more pronounced if you use it to create a view that is queried a lot. Since BigQuery charges are more related to how many columns rather than how many rows are included in the query, SELECT* will select all columns and cost the most. Also, be careful when unnesting columns, since this can also increase the volume of data you're charged for.

There are plenty of SQL examples throughout the book that deal with specific use cases, so this section has been more about the tables' specification that SQL will operate upon. The general principles should help you keep a clean and efficient operation of your BigQuery data that, once adopted, will become a popular tool within your organization.

While BigQuery can deal with streaming data, sometimes event-based data needs a more dedicated tool, which is when Pub/Sub enters the picture.

Pub/Sub

Pub/Sub is integral to how many data imports happen. Pub/Sub is a global messaging system, meaning it's a way to enact the pipes between data sources in an event-driven manner.

Pub/Sub messages have guaranteed at-least-once delivery, so it's a way to ensure consistency in your pipelines. This differs from, say, HTTP API calls, which you shouldn't count on working 100% of the time. Pub/Sub achieves this as the receiving systems must "ack," or acknowledge, that they've received the Pub/Sub message. If it doesn't return an "ack", then Pub/Sub will queue the message to be sent again. This happens at scale—billions of hits can be sent through Pub/Sub; in fact, it's a similar technology to the Googlebot crawler that crawls the entire World Wide Web for Google Search.

Pub/Sub isn't data storage as such, but it does act like the pipes between storage solutions on GCP so is relevant here. Pub/Sub acts like a generic pipeline for you to send data to via its topics, and you can then consume that data at the other end via its subscriptions. You can map many subscriptions to a topic. It can also scale: you can send billions of events through it without worrying about setting up servers, and with its guaranteed at-least-once delivery service, you'll know that they will get through. It can offer this guarantee because each subscription needs to acknowledge that it has received the sent data (or "ack" it, as it's known when talking about message queues), otherwise it will queue it up to send it again.

This topic/subscription model means you can have one event coming in that is sent to several storage applications or event-based triggers. Almost every action on GCP has an option to send a Pub/Sub event, since they can also be triggered via logging filters. This was my first application in using them: BigQuery GA360 exports are notorious for not always coming at the same time each day, which can break downstream import jobs if they're set up on a schedule. Using the log to track when the BigQuery tables were actually populated could then trigger a Pub/Sub event, which could then start the jobs.

Setting Up a Pub/Sub Topic for GA4 BigQuery Exports

A useful Pub/Sub event occurs when your GA4 BigQuery exports are ready, which we can use later for other applications (such as "Cloud Build" on page 137.)

We can do this using the general logs for Google Cloud Console, called Cloud Logging. This is where all logs for all services you're running will sit, including BigQuery. If we can filter down to the services log entries for the activity you want to monitor, you can set up a logs-based metric that will trigger a Pub/Sub topic.

We first need to create a Pub/Sub topic from the Cloud Logging entries that record your BigQuery activity related to when the GA4 export is ready.

Example 4-4 shows an example of a filter for this, with the results in Figure 4-2.

Example 4-4. A filter you can use within Cloud Logging to see when your GA4 BigQuery export is ready

```
resource.type="bigquery_resource"
protoPayload.authenticationInfo.principalEmail=
    "firebase-measurement@system.gserviceaccount.com"
protoPayload.methodName="jobservice.jobcompleted"
```

Applying this filter, we see only the entries when the Firebase service key `firebase-measurement@system.gserviceaccount.com` has finished updating your BigQuery table.

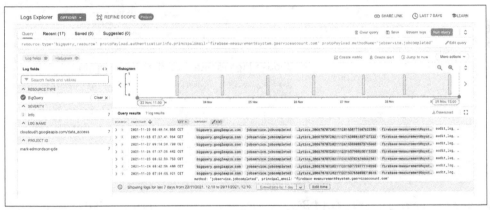

Figure 4-2. A Cloud Logging filter for seeing when your GA4 BigQuery exports are ready, which we can use to create a Pub/Sub topic

Once you're happy with the log filter, select the "Logs Router" to route them into Pub/Sub. An example of the set-up screen is shown in Figure 4-3.

Once the log is created, you should get a Pub/Sub message each time the BigQuery export is ready for consumption later on. I suggest using Cloud Build to process the data, as detailed more in "Cloud Build" on page 137, or following the example in the next section, which will create a BigQuery partitioned table.

Figure 4-3. Setting up your GA4 BigQuery log so it sends the entries to Pub/Sub topic named ga4-bigquery

Creating Partitioned BigQuery Tables from Your GA4 Export

By default, the GA4 exports are in "sharded" tables, which means that each table is created separately and you use wildcards in the SQL to fetch them all, e.g., three days' tables are called `events_20210101`, `events_20210102`, and `events_20210103`, which you can query via the SQL snippet `SELECT * FROM dataset.events_*`—the * is the wildcard.

This works, but if you wish to optimize your downstream queries, then aggregating the tables into one partitioned table will make some jobs flow easier and will allow some query optimizations for speed. We'll use the Pub/Sub topic set up in Figure 4-3 to trigger a job that will copy the table over into a partitioned table.

To do this, go to the Pub/Sub topic and create a Cloud Function to be triggered by it by hitting the button at the top. The code to copy the table over into a partitioned table is in Example 4-5.

Example 4-5. Python code for a Cloud Function to copy your GA4 BigQuery exports into a partitioned table

```python
import logging
import base64
import JSON
from google.cloud import bigquery # pip google-cloud-bigquery==1.5.1
import re

# replace with your dataset
DEST_DATASET = 'REPLACE_DATASET'

def make_partition_tbl_name(table_id):
  t_split = table_id.split('_20')

  name = t_split[0]

  suffix = ''.join(re.findall("\d\d", table_id)[0:4])
  name = name + '$' + suffix

  logging.info('partition table name: {}'.format(name))

  return name


def copy_bq(dataset_id, table_id):
  client = bigquery.Client()
  dest_dataset = DEST_DATASET
  dest_table = make_partition_tbl_name(table_id)

  source_table_ref = client.dataset(dataset_id).table(table_id)
  dest_table_ref = client.dataset(dest_dataset).table(dest_table)

  job = client.copy_table(
    source_table_ref,
    dest_table_ref,
    location = 'EU') # API request

  logging.info(f"Copy job:
    dataset {dataset_id}: tableId {table_id} ->
    dataset {dest_dataset}: tableId {dest_table} -
    check BigQuery logs of job_id: {job.job_id}
    for status")

def extract_data(data):
  """Gets the tableId, datasetId from pub/sub data"""
  data = JSON.loads(data)
```

```
    complete = data['protoPayload']['serviceData']['jobCompletedEvent']['job']
    table_info = complete['jobConfiguration']['load']['destinationTable']
    logging.info('Found data: {}'.format(JSON.dumps(table_info)))
    return table_info

def bq_to_bq(data, context):
  if 'data' in data:
    table_info = extract_data(base64.b64decode(data['data']).decode('utf-8'))
    copy_bq(dataset_id=table_info['datasetId'], table_id=table_info['tableId'])
  else:
    raise ValueError('No data found in pub-sub')
```

Deploy the Cloud Function with its own service account, and give that service account BigQuery Data Owner permissions. If possible, try to restrict to as specific a dataset or table as you can as a best practice.

Once the Cloud Function is deployed, your GA4 BigQuery exports will be duplicated into a partitioned table in another dataset. The Cloud Function reacts to the Pub/Sub message that the GA4 export is ready and triggers a BigQuery job to copy the table. This is helpful for applications such as the Data Loss Prevention API, which does not work with sharded tables, and is shown in an example application in "Data Loss Prevention API" on page 159.

Server-side Push to Pub/Sub

Another use for Pub/Sub is as part of your data collection pipeline if using GTM SS. From your GTM SS container, you can push all the event data to a Pub/Sub endpoint for use later.

Within GTM SS, you can create a container that will send all the event data to an HTTP endpoint. That HTTP endpoint can be a Cloud Function that will transfer it to a Pub/Sub topic—code to do that is shown in Example 4-6.

Example 4-6. Some example code to show how to send the GTM SS events to an HTTP endpoint, which will convert it into a Pub/Sub topic

```
const getAllEventData = require('getAllEventData');
const log = require("logToConsole");
const JSON = require("JSON");
const sendHttpRequest = require('sendHttpRequest');

log(data);

const postBody = JSON.stringify(getAllEventData());

log('postBody parsed to:', postBody);

const url = data.endpoint + '/' + data.topic_path;
```

```
log('Sending event data to:' + url);

const options = {method: 'POST',
        headers: {'Content-Type':'application/JSON'}};

// Sends a POST request
sendHttpRequest(url, (statusCode) => {
 if (statusCode >= 200 && statusCode < 300) {
  data.gtmOnSuccess();
 } else {
  data.gtmOnFailure();
 }
}, options, postBody);
```

A Cloud Function can be deployed to receive this HTTP endpoint with the GTM SS event payload and create a Pub/Sub topic, as shown in Example 4-7.

Example 4-7. An HTTP Cloud Function pointed to within the GTM SS tag that will retrieve the GTM SS event data and create a Pub/Sub topic with its content

```
import os, JSON
from google.cloud import pubsub_v1 # google-cloud-Pub/Sub==2.8.0

def http_to_Pub/Sub(request):
  request_JSON = request.get_JSON()
  request_args = request.args

  print('Request JSON: {}'.format(request_JSON))

  if request_JSON:
    res = trigger(JSON.dumps(request_JSON).encode('utf-8'), request.path)
    return res
  else:
    return 'No data found', 204


def trigger(data, topic_name):
 publisher = Pub/Sub_v1.PublisherClient()

 project_id = os.getenv('GCP_PROJECT')
 topic_name = f"projects/{project_id}/topics/{topic_name}"

 print ('Publishing message to topic {}'.format(topic_name))

 # create topic if necessary
 try:
  future = publisher.publish(topic_name, data)
  future_return = future.result()
  print('Published message {}'.format(future_return))

  return future_return
```

```
except Exception as e:
  print('Topic {} does not exist? Attempting to create it'.format(topic_name))
  print('Error: {}'.format(e))

  publisher.create_topic(name=topic_name)
  print ('Topic created ' + topic_name)

  return 'Topic Created', 201
```

Firestore

Firestore is a NoSQL database as opposed to the SQL you may use in products such as "BigQuery" on page 109. As a complement for BigQuery, Firestore (or Datastore) is a counterpart that focuses on fast response times. Firestore works via keys that are used for quick lookups of data associated with it—and by *quick*, we mean subsecond. This means you must work with it in a different way than with BigQuery; most of the time, requests to the database should refer to a key (like a user ID) that returns an object (such as user properties).

 Firestore used to be called Datastore and is a rebranding of the product. Taking the best of Datastore and another product called Firebase Realtime Database, Firestore is a NoSQL document database built for automatic scaling, high performance, and ease of application development.

Firestore is linked to the Firebase suite of products and is usually used for mobile applications that need first lookups with mobile support via caching, batching, etc. Its properties can also be helpful for analytics applications because it's ideal for fast lookups when giving an ID, such as a user ID.

When to Use Firestore

I typically use Firestore when I'm looking at creating APIs that will possibly be called multiple times per second, such as serving up a user's attributes when given their user ID. This is usually more to support the data activation end of a project, with a light API that will take your ID, query the Firestore, and return with the attributes all within a few microseconds.

If you ever need a fast lookup, then Firestore will also be handy. An example that is powerful for analytics tracking is to keep your products database in a products Firestore with a lookup on the product SKU that returns that product's cost, brand, category, etc. With such a database in place, you can improve your analytics collection by trimming down the ecommerce hits to include only the SKU and look up the data before sending it to GA4. This allows you to send much smaller hits from the user's web browser with security, speed, and efficiency benefits.

Accessing Firestore Data Via an API

To access Firestore, you first need to import your data into a Firestore instance. You can do this via its import APIs or even by manually inputting via the WebUI. The requirement of the dataset is that you will always have a key that will be typically what you send to the database to return data, and then a nested JSON structure of data will come back.

Adding data to Firestore involves defining the object you want to record that could be in a nested structure and its location in the database. Altogether this defines a Firestore document. An example of how this would be added via Python is shown in Example 4-8.

Example 4-8. Importing a data structure into Firestore using the Python SDK, in this case, a demo product SKU with some details

```
from google.cloud import firestore
db = firestore.Client()

product_id = u'SKU12345'

data = {
  u'name': u'Muffins',
  u'brand': u'Mule',
  u'price': 15.78
}

# Add a new doc in collection 'your-firestore-collection'
db.collection(u'your-firestore-collection').document(product_id).set(data)
```

Using this means that you may need an additional data pipeline for importing your data into Firebase so that you can look up the data from your applications, which would use code similar to that in Example 4-8.

Once you have your data in Firestore, you can then reach it via your application. Example 4-9 gives a Python function you can use in a Cloud Function or App Engine application. We assume it's being used to look up product information when it's supplied a `product_id`.

Example 4-9. An example of how to read data from a Firestore database using Python within a Cloud Function

```
# pip google-cloud-firestore==2.3.4
from google.cloud import firestore

def read_firestore(product_id):
 db = firestore.Client()
 fs = 'your-firestore-collection'
 try:
  doc_ref = db.collection(fs).document(product_id)
 except:
  print(f'Could not connect to firestore collection: {fs}')
  return {}

 doc = doc_ref.get()
 if doc.exists:
  print(f'product_id data found: {doc.to_dict()}')
  return doc.to_dict()
 else:
  print(f'Could not find entry for product_id: {product_id}')
  return {}
```

Firestore gives you another tool that can help your digital analytics workflows and will come more to the fore when you need real-time applications and millisecond response times, such as calling from an API or as a user browses your website and you don't want to add latency to their journey. It's more suited to web application frameworks than to data analysis tasks, so it's often used during the last steps of data activation.

BigQuery and Firestore are both examples of databases that work with structured data, but you'll also come across unstructured data such as videos, pictures, or audio or just data you don't know the shape of before you process it. In that case, your storage options need to work on a more low level of storing bytes, and that is where Cloud Storage comes into the picture.

GCS

We've already talked about using GCS for ingesting data into CRM systems in "Google Cloud Storage" on page 82, but this section is more about its use generally. GCS is useful for several roles, helped by the simple task it excels at: keeping bytes secure but instantly available.

GCS is the GCP service storage system most like the hard drive storing the files sitting on your computer. You can't manipulate or do anything to that data until you

---

open it up in an application, but it will store TBs of data for you to access in a secure and accessible manner. The roles I use it for are the following:

Unstructured data

For objects that can't be loaded into a database such as video and images, GCS is a location that will always be able to help. It can store anything within bytes in its buckets, objects that are affectionately known as "blobs." When working with Google APIs such as speech-to-text or image recognition, the files usually need to be uploaded to GCS first.

Raw data backups

Even for structured data, GCS is helpful as a raw data backup that can be stored at its archival low rates so you can always rewind or disaster-recover your way back from an outage.

Data import landing pads

As seen in "Google Cloud Storage" on page 82, GCS is helpful as a landing pad for export data since it won't be fussy about the data schema or format. Since it will also trigger Pub/Sub events when data does arrive, it can start the event-based data flow systems.

Hosting websites

You can choose to make files publicly available from HTTP endpoints, meaning if you place HTML or other files supported by web browsers, you can have static websites hosted in GCS. This can also be helpful for static assets you may wish to import into websites such as for tracking pixels or images.

Dropbox

You can give public or more fine-grained access to certain users so you can securely pass on large files. Up to 5 TBs per object is supported, with unlimited (if you're prepared to pay!) overall storage. This makes it a potential destination for data processing, such as a CSV file made available to colleagues who wish to import it locally into Excel.

Items stored in GCS are all stored at their own URI, which is like an HTTP address (https://example.com) but with its own protocol: gs://. You can also make them available at a normal HTTP address—in fact, you could host HTML files and GCS would serve as your web hosting.

The bucket names you use are globally unique, so you can access them from any project even if that bucket sits in another. You can specify public access over HTTP or only specific users or service emails working on behalf of your data applications. Figure 4-4 shows an example of how this looks via the WebUI, but the files within are usually accessed via code.

Figure 4-4. Files sitting within GCS in its WebUI

Each object in GCS has some metadata associated with it that you can use to tailor it to your storage needs. We'll walk through the example file shown in Figure 4-5 to help illustrate what is possible.

Figure 4-5. Various metadata associated with a file upload to GCS

The metadata available for each object within GCS includes:

Type
This is an HTTP MIME (Multipurpose Internet Mail Extensions) type as specified for web objects. Mozilla's website (*https://oreil.ly/TxUdt*) has some resources on HTTP MIME types. It's worth setting this if your application will check against it to determine how to treat the file—for instance, a `.csv` file with MIME type `text/csv` in Figure 4-5 means applications downloading it will attempt to read it as a table. Other common MIME types you may come across are JSON (`application/JSON`), HTML for web pages (`text/html`), images such as `image/png`, and video (`video/mp4`).

Size
The size on disk of the object's bytes. You can store up to 5 TB per object.

Created
When the object was first created.

Last modified
You can update objects by calling them the same name as when you first created them, and have object versioning activated.

Storage class
The pricing model that object is stored under, set at the bucket level. The storage classes are generally a compromise between storage cost and access cost. The costs for storage vary per region, but as a guide, here are some examples for GBs per month. *Standard* is for data that is accessed frequently ($0.02), *Nearline* for data that may be accessed a only few times a year ($0.01), *Coldline* for data that may be accessed only annually or less ($0.004), and *Archive* for data that may never be accessed aside from disaster recovery ($0.0012). Make sure to put your objects in the right class, or you will end up overpaying for data access because object cost prices are higher for accessing Archive data than for Standard, for example.

Custom time
You may have important dates or times to associate with the object, which you can add as metadata here.

Public URL
If you choose to make your object public, the URL will be listed here. Note that this is different from the Authenticated URL.

Authenticated URL
This is the URL if you're giving restricted, not public, access to a user or application. It will check against that user's authorization before serving up the object.

gstuil URI

The `gs://` form of accessing the object, most typically when using it program-matically via the API or one of GCS's SDKs.

Permission

Information about who can access the object. It's typical these days for permission to be granted on a bucket level, although you can also choose to fine-grain control objects access. It's usually easier to have two separate buckets for access control, such as public and restricted.

Protection

There are various methods you can enable to control how the object persists, which are highlighted in this section.

Hold status

You can enforce temporary or event-based holds on the object, meaning it can't be deleted or modified when in place, either by a time limit or when a certain event triggered by an API call happens. This can be helpful to protect against accidental deletion, or if, for example, you have a data expiration active on the bucket via a retention policy but want to keep certain objects out of that policy.

Version history

You can enable versioning on your object such that even if it is modified, the older version will still be accessible. This can be helpful to keep a track record of scheduled data.

Retention policy

You can enable various rules that determine how long an object sticks around. This is vital if you're dealing with personal user data to delete old archives once you no longer have permission to hold that data. You can also use it to move data into a lower-cost storage solution if it isn't accessed after a certain number of days.

Encryption type

By default, Google operates an encryption approach to all your data on GCP, but you may want to enforce a tighter security policy in which not even Google can see the data. You cam do this by using your own security keys.

GCS is singular in purpose, but it's a fundamental purpose: store your bytes safely and securely. It is the bedrock that several other GCP services rely on, even if they're not exposed to the end user and can serve that role for you too. It's an infinite hard drive in the cloud rather than on your own computer, and it can be easily accessed across the world.

We've now looked at the three major data storage types: BigQuery for structured SQL data, Firestore for NoSQL data, and GCS for unstructured raw data. We now turn to how you work with them on a regular basis, looking at techniques for scheduling and streaming data flows. Let's start with the most common application, scheduled flows.

Scheduling Data Imports

This section looks at one of the major tasks for any data engineer designing workflows: how to schedule data flows within your applications. Once your proof of concept is working, your next step to putting that into production is being able to regularly update the data involved. Rather than updating a spreadsheet each day or running an API script, turning this task over to the many automation devices available to you on GCP ensures that you will have continuously updated data without needing to worry about it.

There are many ways to approach data updates, which this section will go through in relation to how you want to move your GA4 data and its companion datasets.

Data Import Types: Streaming Versus Scheduled Batches

Streaming data versus batching data is one of those decisions that you may come across when designing data application systems. This section considers some of the advantages and disadvantages of both.

Streaming data flows are more real time, using event-based small data packets that are continuously being updated. Batched data is regularly scheduled at a slower interval, such as daily or hourly, with larger data imports each job.

The streaming options for data will be considered more in depth in "Streaming Data Flows" on page 145, but comparing them side by side with batched data can help with some fundamental decisions early on in your application design.

Batched data flows

Batching is the most common and traditional way of importing data flows, and for most use cases it's perfectly adequate. A key question when creating your use case will be how fast you need that data. It's common for the initial reaction to be as fast as possible or near real time. But looking at the specifics, you find that actually the effects of hourly or even daily updates will be unnoticeable compared to real time, and these types of updates will be a lot cheaper and easier to run. If the data you're updating with is also batched (for example, a CRM export that happens nightly), there will be little reason to make downstream data real time. As always, look at the use case application and see if it makes sense. Batched data workflows start to break down if you can't rely on those scheduled updates to be on time. You may then need to create fallback options if an import fails (and you should always design for eventual failure).

Streaming data flows

Streaming data is easier to do these days in modern data stacks given the new technologies available, and there are proponents who will say all your data flows should be streaming if possible. It may well be that you discover new use cases once you break free from the shackles of batched data schedules. There are certain advantages even if you're not in immediate need of real-time data since when moving to an event-based model of data we react when something happens, not when a certain timestamp is reached, which means we can be more flexible in when data flows occur. A good example of this is the GA4 BigQuery data exports, which if delayed would break downstream dashboards and applications. Setting up event-based reactions to when the data is available means you will get the data as soon as it's there, rather than having to wait for the next day's delivery. The major disadvantage is cost because these flows are typically more expensive to run. Your data engineers will also need a different skill level to be able to develop and troubleshoot streaming pipelines.

When considering scheduled jobs, we'll start with BigQuery's own resources, then branch out into more sophisticated solutions such as Cloud Composer, Cloud Scheduler, and Cloud Build.

BigQuery Views

In some cases, the simplest way to present transformed data is to set up a BigQuery View or schedule BigQuery SQL. This is the easiest to set up and involves no other services.

BigQuery Views are not tables in the traditional sense but rather represent a table that would result from the SQL you use to define it. This means that when you create your SQL, you can include dynamic dates and thus always have the latest data. For example, you could be querying your GA4 BigQuery data exports with a View created as in Example 4-10—this will always bring yesterday's data back.

Example 4-10. This SQL can be used in a BigQuery View to always show yesterday's data (adapted from Example 3-6)

```
SELECT
  -- event_date (the date on which the event was logged)
  parse_date('%Y%m%d',event_date) as event_date,
  -- event_timestamp (in microseconds, utc)
  timestamp_micros(event_timestamp) as event_timestamp,
  -- event_name (the name of the event)
  event_name,
  -- event_key (the event parameter's key)
  (SELECT key FROM UNNEST(event_params)
   WHERE key = 'page_location') as event_key,
  -- event_string_value (the string value of the event parameter)
```

```
    (SELECT value.string_value FROM UNNEST(event_params)
      WHERE key = 'page_location') as event_string_value
FROM
    -- your GA4 exports - change to your location
    `learning-ga4.analytics_250021309.events_*`
WHERE
    -- limits query to use table from yesterday only
    _TABLE_SUFFIX = FORMAT_DATE('%Y%m%d',date_sub(current_date(), INTERVAL 1 day))
    -- limits query to only show this event
    and event_name = 'page_view'
```

The key line is `FORMAT_DATE('%Y%m%d',date_sub(current_date(), INTERVAL 1 day))`, which returns yesterday, which takes advantage of the `_TABLE_SUFFIX` column BigQuery adds as meta information about the table so that you can more easily query multiple tables.

BigQuery Views have their place, but be careful using them. Since the View SQL is running underneath any other queries run against them, you can run into expensive or slow results. This has been recently mitigated with Materialized Views, which is a technology to make sure you don't query against the entire table when making queries on top of Views. In some cases, you may be better off creating your own intermediate table, perhaps via a scheduler to create the table, which we cover in the next section.

BigQuery Scheduled Queries

BigQuery has native support for scheduling queries, accessible via the menu bar at the top left or by selecting "Schedule" when you create the query. This is fine for small jobs and imports; I would, however, caution against relying on this for anything other than single-step, simple transformations. Once you're looking at more complicated data flows, then it will become easier to use dedicated tools for the job, both from a management and robustness perspective.

Scheduled queries are tied to the user authentication who sets them up, so if that person leaves, the scheduler will need to be updated via the gcloud command `bq update --transfer-config --update-credentials`. Perhaps use this to update your connection to service accounts that are not linked to one person. You'll also have only the BigQuery Scheduler interface to control the queries—for large complicated queries you want to modify, this will make it hard to see a change history or overview.

But for simple, nonbusiness-critical queries that are needed perhaps for a limited number of people, it is quick and easy to set up within the interface itself, and it would serve better than Views for, say, exports to dashboard solutions such as Looker or Data Studio. As seen in Figure 4-6, once you have developed your SQL and have results you like, you can hit the "Schedule" button and have the data ready for you when you log in the next day.

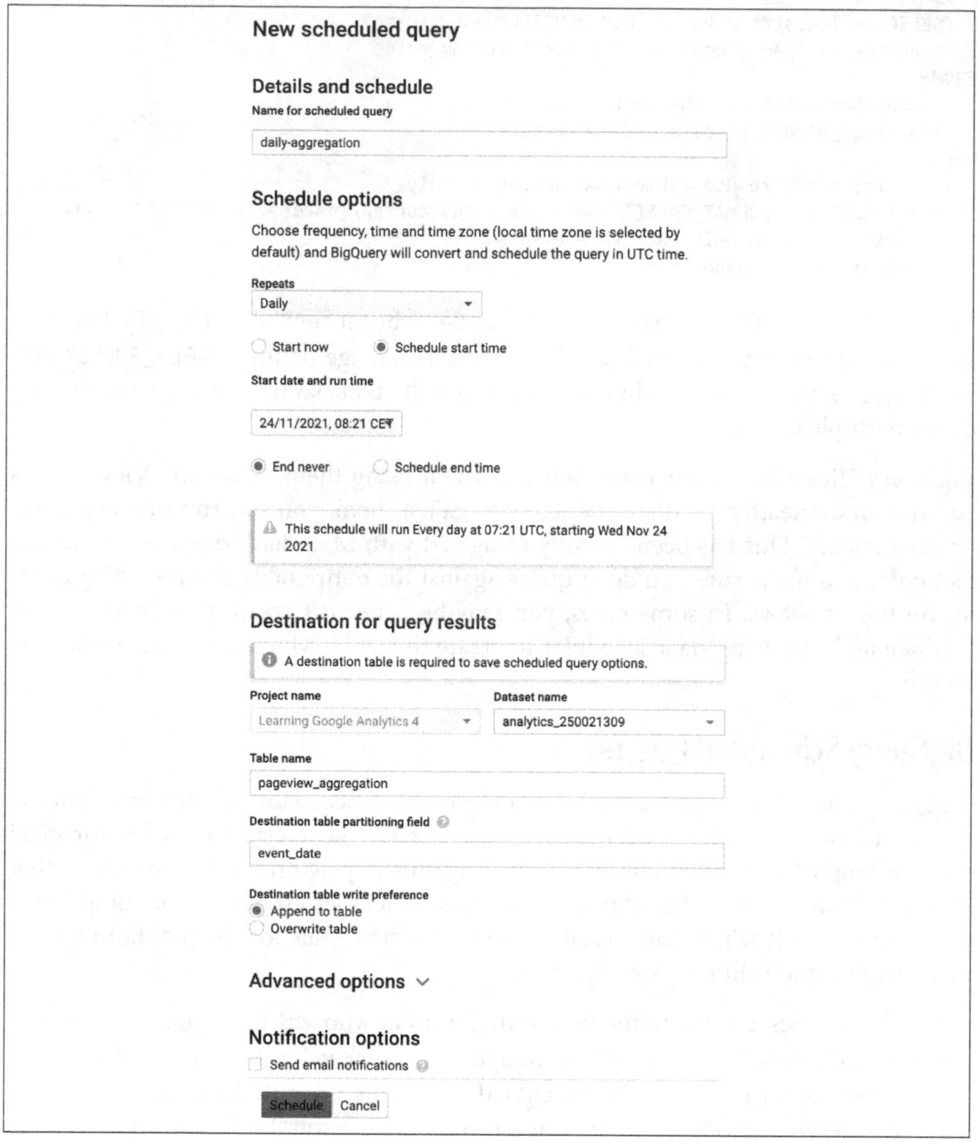

New scheduled query

Details and schedule

Name for scheduled query

> daily-aggregation

Schedule options

Choose frequency, time and time zone (local time zone is selected by default) and BigQuery will convert and schedule the query in UTC time.

Repeats

> Daily ▼

○ Start now   ● Schedule start time

Start date and run time

> 24/11/2021, 08:21 CET

● End never   ○ Schedule end time

> ⚠ This schedule will run Every day at 07:21 UTC, starting Wed Nov 24 2021

Destination for query results

> ⓘ A destination table is required to save scheduled query options.

Project name   Dataset name

> Learning Google Analytics 4 ▼   analytics_250021309 ▼

Table name

> pageview_aggregation

Destination table partitioning field ⓘ

> event_date

Destination table write preference
● Append to table
○ Overwrite table

Advanced options ⌄

Notification options
☐ Send email notifications ⓘ

[Schedule] Cancel

Figure 4-6. Setting up a scheduled query from Example 4-10, which may perform better than creating a BigQuery View with the same data for use in dashboards, etc.

However, once you start asking questions like, "How can I make this scheduled query more robust?" or "How can I trigger off queries based on the data I'm creating in this table?" this is a sign you will need a more robust solution for scheduling. The tool for that job is Airflow, via its hosted version on GCP called Cloud Composer that we talk about in the next section.

Cloud Composer

Cloud Composer is a Google-managed solution for Airflow (*https://oreil.ly/hCqBX*), a popular open source scheduling tool. It costs around $300 a month, so it's only worth looking at once you have some good business value to justify it, but it is the solution I trust the most when looking at complicated data flows across multiple systems, offering backfilling, alerting systems, and configuration via Python. Many companies consider it the backbone of all of their scheduling jobs.

 I will use the name Cloud Composer in this book since that is what managed Airflow is called within the GCP, but a large amount of the content will be applicable to Airflow running on other platforms as well, such as other cloud providers or self-hosted platform.

I started to use Cloud Composer once I had jobs that fulfilled the following criteria:

Multilevel dependencies
> As soon as you have a situation in your data pipelines where one scheduled job depends on another, I would start using Cloud Composer since it will fit well within its directed acyclic graph (DAG) structure. Examples include chains of SQL jobs: one SQL script to tidy the data, another SQL script to make your model data. Putting these SQL scripts to be executed within Cloud Composer allows you to break up your scheduled jobs into smaller, simpler components than if try to run them all in one bigger job. Once you have the freedom to set dependencies, I recommend improving pipelines by adding checks and validation steps that would be too complex to do via a single scheduled job.

Backfills
> It's common to set up a historic import at the start of the project to backfill all the data you would have had if the schedule job was running, for example, the past 12 months. What is available will differ per job, but if you have set up imports per day it can sometimes be nontrivial to then set up historic imports. Cloud Composer runs jobs as a simulation of the day, and you can set any start date so it will slowly backfill all data if you let it.

Multiple interaction systems
> If you're pulling in data or sending data to multiple systems such as FTP, Cloud products, SQL databases, and APIs, it will start to become complex to coordinate across those systems and may require spreading them out across various import scripts. Cloud Composer's many connectors via its Operators and Hooks means it can connect to virtually anything, so you can manage them all from one place, which is a lot easier to maintain.

Retries

As you import over HTTP, in a lot of cases, you will experience outages. It can be difficult to configure when and how often to retry those imports or exports, which Cloud Composer can help with via its configurable retry system that controls each task.

Once working with data flows, you will quickly experience issues like the ones mentioned and need a way to solve them easily, for which Cloud Composer is one solution. Similar solutions exist, but Cloud Composer is the one I've used the most and has quickly become the backbone of many data projects. How to represent these flows in an intuitive manner is useful to be able to imagine complicated processes, which Cloud Composer solves using a representation we talk about next.

DAGs

The central feature of Cloud Composer are DAGs, which represent the flow of your data as it is ingested, processed, and extracted. The name refers to a node and edge structure, with directions between those nodes given by arrows. An example of what that could mean for your own GA4 pipeline is shown in Figure 4-7.

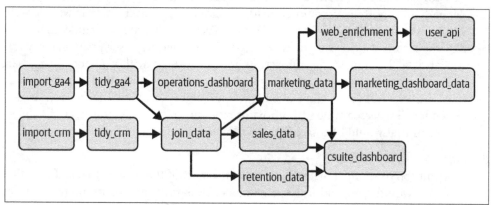

Figure 4-7. An example of a DAG that could be used in a GA4 process

The nodes represent a data operation, with the edges showing the order of events and which operations are dependent on one another. One of the key features of Airflow is that if one node fails (which eventually all do), then it has strategies in place to either wait, retry, or skip downstream operations. It also has some backfill features that can prevent a lot of headaches from running historic updates, and it comes with some predefined macros that allow you to dynamically insert, for example, today's date into your scripts.

An example of a DAG that imports from your GA4 BigQuery exports is shown in Example 4-11.

Example 4-11. An example DAG that takes your GA4 export and aggregates it using SQL that you have developed earlier and in a ga4-bigquery.sql file uploaded with your script

```
from airflow.contrib.operators.bigquery_operator import BigQueryOperator
from airflow.contrib.operators.bigquery_check_operator import BigQueryCheckOperator
from airflow.operators.dummy_operator import DummyOperator
from airflow import DAG
from airflow.utils.dates import days_ago
import datetime

VERSION = '0.1.7' # increment this each version of the DAG

DAG_NAME = 'ga4-transformation-' + VERSION

default_args = {
  'start_date': days_ago(1), # change this to a fixed date for backfilling
  'email_on_failure': True,
  'email': 'mark@example.com',
  'email_on_retry': False,
  'depends_on_past': False,
  'retries': 3,
  'retry_delay': datetime.timedelta(minutes=10),
  'project_id': 'learning-ga4',
  'execution_timeout': datetime.timedelta(minutes=60)
}

schedule_interval = '2 4 * * *' # min, hour, day of month, month, day of week

dag = DAG(DAG_NAME, default_args=default_args, schedule_interval=schedule_interval)


start = DummyOperator(
  task_id='start',
  dag=dag
)

# uses the Airflow macro {{ ds_nodash }} to insert todays date in YYYYMMDD form
check_table = BigQueryCheckOperator(
  task_id='check_table',
  dag=dag,
  sql='''
  SELECT count(1) > 5000
  FROM `learning-ga4.analytics_250021309.events_{{ ds_nodash }}`"
  '''
)

checked = DummyOperator(
  task_id='checked',
  dag=dag
)
```

```
# a function so you can loop over many tables, SQL files
def make_bq(table_id):

  task = BigQueryOperator(
    task_id='make_bq_'+table_id,
    write_disposition='WRITE_TRUNCATE',
    create_disposition='CREATE_IF_NEEDED',
    destination_dataset_table=
        'learning_ga4.ga4_aggregations.{}${{ ds_nodash}}'.format(table_id),
    sql='./ga4_sql/{}.sql'.format(table_id),
    use_legacy_sql=False,
    dag=dag
  )

  return task

ga_tables = [
 'pageview-aggs',
 'ga4-join-crm',
 'ecom-fields'
]

ga_aggregations = [] # helpful if you are doing other downstream transformations
for table in ga_tables:
 task = make_bq(table)
 checked >> task
 ga_aggregations.append(task)


# create the DAG
start >> check_table >> checked
```

To make the nodes for your DAG, you will use Airflow Operators, which are various functions premade to connect to a variety of applications, including an extensive GCP array of services such as BigQuery, FTP, Kubernetes clusters, and so on.

For the example in Example 4-11, the nodes are created by:

start

A DummyOperator() to signpost the start of the DAG.

check_table

A BigQueryCheckOperator() that will check that you have data that day in the GA4 table. If this fails by returning FALSE for the inline SQL shown, Airflow will fail the task and retry it again every 10 minutes up to 3 times. You can modify this to your expectations.

checked

Another DummyOperator() to signpost that the table has been checked.

---

make_bq

This will create or add to a partitioned table with the same name as the task_id. The SQL it will execute should also have the same name and be available in the SQL folder uploaded with the DAG, in `./ga4_sql/`, e.g., `./ga4_sql/pageview-aggs.sql`. It is functionalized so you can loop over `tableIds` for more efficient code.

The edges are taken care of via bitwise Python operators at the end of the tag and within the loops, e.g., `start >> check_table >> checked`.

You can see the resulting DAG displayed in Figure 4-8. Take this example as a basis you can expand for your own workflows.

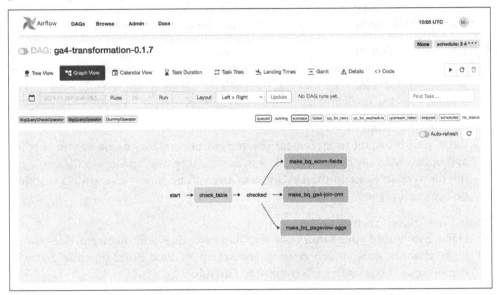

Figure 4-8. An example of the DAG created in Airflow by the code in Example 4-11; to scale up for more transformations, add more SQL files to the folder and add the name of the table to the ga_tables list

Tips for using Airflow/Cloud Composer

The general help files are excellent for learning how to use Cloud Composer, but the following are some tips I've picked up while using it within data science projects:

Use Airflow only for scheduling

Use the right tool for the right job—Airflow's role is scheduling and connecting to data storage systems. I made the mistake of using its Python libraries to massage data a bit between scheduling steps but got into a Python dependency hell that affected all running tasks. I prefer to use Docker containers for any code that needs to run, and use the `GKEPodOperator()` instead to run that code in a controlled environment.

Write functions for your DAGS
It's much cleaner to create functions that output DAGs rather than having the tasks written out each time. It also means you can loop over them and create dependencies for lots of datasets at once without needing to copy-paste code.

Use dummy operators to signpost
The DAGs look impressive but can be confusing, so having some handy signposts along the line can indicate where you can stop and start misbehaving DAG runs. Being able to clear all downstream from "Data all loaded" signpost makes it clear what will happen. Other features that can help here are Task Groups and Labels, which can help display meta information about what your DAG is doing.

Separate out your SQL files
You don't have to write out huge strings of SQL for your operators; you can instead put them into *.sql* files and then call the file holding the SQL. This makes it much easier to track and keep on top of changes.

Version your DAG names
I also find it helpful to increment the version of the DAG name as you modify and update. Airflow can be a bit slow in recognizing new updates to files, so having the version name in the DAG means you can be sure you're always working on the latest version.

Set up Cloud Build to deploy DAGs
Having to upload your DAG code and files each time will disincentivize you to make changes, so it's much easier if you set up a Cloud Build pipeline that will deploy your DAGs upon each commit to GitHub.

That was a flying tour of Cloud Composer features, but there is a lot more to it, and I recommend the Airflow website to explore more of its options. It is a heavyweight scheduling option, and there is another, much lighter, option in Google Cloud Scheduler, which we'll look at next.

Cloud Scheduler

If you're looking for something more lightweight than Cloud Composer, then Cloud Scheduler is a simple cron-in-the-cloud service you can use to trigger HTTP endpoints. For simple tasks that don't need the complexity of data flows supported by Cloud Composer, it just works.

I put it somewhere in between Cloud Composer and BigQuery-scheduled queries in capabilities, since Cloud Scheduler will run not just BigQuery queries but any other GCP service as well since that can be handy.

To do that, it does involve some extra work to create the Pub/Sub topic and Cloud Function that will create the BigQuery job, so if it's just BigQuery, you may not have a need, but if other GCP services are involved, centralizing the location of your scheduling may be better in the long run. You can see an example of setting up a Pub/Sub topic in "Pub/Sub" on page 114; the only difference is that then you schedule an event to hit that topic via Cloud Scheduler. You can see some examples from my own GCP in Figure 4-9 that shows the following:

Packagetest-build
    A weekly schedule to trigger an API call to run a Cloud Build

Slackbot-schedule
    A weekly schedule to hit an HTTP endpoint that will trigger a Slackbot

Target_Pub/Sub_scheduler
    A daily schedule to trigger a Pub/Sub topic

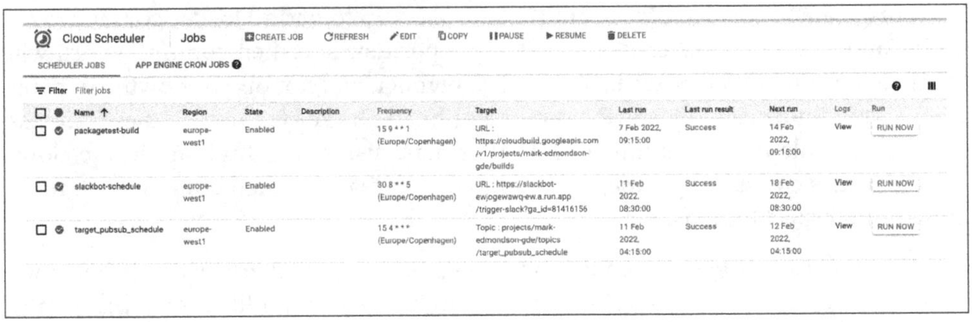

Figure 4-9. Some Cloud Schedules I enabled for some tasks within my own GCP (https://oreil.ly/QGObe)

Cloud Scheduler can also trigger other services such as Cloud Run or Cloud Build. A particularly powerful combination is Cloud Scheduler and Cloud Build (covered in the next section, "Cloud Build" on page 137). Since Cloud Build can run long-running tasks, you have an easy way to create a serverless system that can run any job on GCP, all event-driven but with some scheduling on top.

Cloud Build

Cloud Build (*https://cloud.google.com/build*) is a powerful tool to consider for data workflows, and it's probably the tool I use most every day (even more than Big-Query!). Cloud Build was also introduced in the data ingestion section in "Setting Up Cloud Build CI/CD with GitHub" on page 95, but we'll go into more detail here.

Cloud Build is classified as a CI/CD tool, which is a popular strategy in modern-day data ops. It mandates that code releases to production should not be at the end of massive development times but with little updates all the time, with automatic testing and deployment features so that any errors are quickly discovered and can be rolled back. These are good practices in general, and I encourage reading up on how to follow them (*https://oreil.ly/AoIZJ*). Cloud Build can also be thought of as a generic way to trigger any code on a compute cluster in reaction to events. The primary intention is for when you commit code to a Git repository such as GitHub, but those events could also be when a file hits GCS, a Pub/Sub message is sent, or a scheduler pings the endpoint.

Cloud Build works by you defining the sequences of events, much like an Airflow DAG but with a simpler structure. For each step, you define a Docker environment to run your code within, and the results of that code running can be passed on to subsequent steps or archived on GCS. Since it works with any Docker container, you can run a multitude of different code environments on the same data; for instance, one step could be Python to read from an API, then R to parse it out, then Go to send the result someplace else.

I was originally introduced to Cloud Build as the way to build Docker containers on GCP. You place your Dockerfile in a GitHub repository and then commit, which will trigger a job and build Docker in a serverless manner, and not on your own computer as you would normally. This is the only way I build Docker containers these days, as building them locally takes time and a lot of hard disk space. Building in the cloud usually means committing code, going for a cup of tea, and then coming back in 10 minutes to inspect the logs.

Cloud Build has now been extended to building not only Dockerfiles but also its own YAML configuration syntax (cloudbuild.yaml) as well as buildpacks. This greatly extends its utility because, via the same actions (Git commit, Pub/Sub event, or a schedule), you can trigger off jobs to do a whole variety of useful tasks, not just Docker containers but running any code you require.

I've distilled the lessons I've learned from working with Cloud Build along with the HTTP Docker equivalent Cloud Run and Cloud Scheduler into my R package `google CloudRunner` (*https://oreil.ly/ELw9m*), which is the tool I use to deploy most of my data engineering tasks for GA4 and otherwise on GCP. Cloud Build uses Docker containers to run everything. I can run almost any language/program or application, including R. Having an easy way to create and trigger these builds from R means R can serve as a UI or gateway to any other program, e.g., R can trigger a Cloud Build using `gcloud` to deploy Cloud Run applications.

Cloud Build configurations

As a quick introduction, a Cloud Build YAML file looks something like that shown in Example 4-12. The example shows how three different Docker containers can be used within the same build, doing different things but working on the same data.

Example 4-12. An example of a cloudbuild.yaml *file used to create builds. Each step happens sequentially. The* name *field is of a Docker image that will run the command specified in the* args *field.*

```
steps:
- name: 'gcr.io/cloud-builders/docker'
  id: Docker Version
  args: ["version"]
- name: 'alpine'
  id: Hello Cloud Build
  args: ["echo", "Hello Cloud Build"]
- name: 'rocker/r-base'
  id: Hello R
  args: ["R", "-e", "paste0('1 + 1 = ', 1+1)"]
```

You then submit this build using the GCP web console, gcloud or googleCloudRunner, or otherwise using the Cloud Build API. The gcloud version is gcloud builds submit --config cloudbuild.yaml --no-source. This will trigger a build in the console where you can watch it via its logs or otherwise—see Figure 4-10 of an example for the googleCloudRunner package checks:

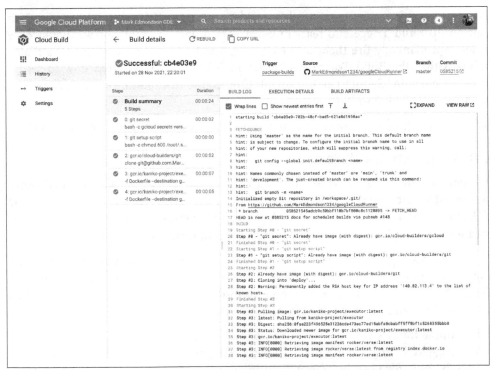

Figure 4-10. A Cloud Build that has successfully built within the Google Cloud Console

We've also seen Cloud Build used to deploy a Cloud Function in "Setting Up Cloud Build CI/CD with GitHub" on page 95—that example is replicated in Example 4-13. Only one step is needed to deploy the Cloud Function from Example 3-9.

Example 4-13. Cloud Build YAML for deploying a Cloud Function from Example 3-9

```
steps:
- name: gcr.io/cloud-builders/gcloud
  args: ['functions',
      'deploy',
      'gcs_to_bq',
      '--runtime=python39',
      '--region=europe-west1',
      '--trigger-resource=marks-crm-imports-2021',
      '--trigger-event=google.storage.object.finalize']
```

Builds can be triggered manually, but often you want this to be an automatic process, which starts to embrace the CI philosophy. For those, you use Build Triggers.

Build Triggers

Build Triggers is a configuration that decides when your Cloud Build will fire. You can set up Build Triggers to react to Git pushes, Pub/Sub events, or webhooks only when you manually fire them in the console. The build can be specified in a file or inline within the Build Trigger configuration. We've already covered how to set up Build Triggers in "Setting Up the GitHub Connection to Cloud Build" on page 95, so see that section for a walk-through.

We covered Cloud Build in general, but we now move to a specific example for GA4.

GA4 applications for Cloud Build

In general, I deploy all code for working with GA4 data through Cloud Build, since it's linked to the GitHub repository I put my code within when not working in the GA4 interface. That covers Airflow DAGs, Cloud Functions, BigQuery tables, etc., via various Cloud Build steps invoking gcloud, my R libraries, or otherwise.

When processing the standard GA4 BigQuery exports, Cloud Logging creates an entry showing when those tables are ready, which you can then use to create a Pub/Sub message. This can kick off an event-driven data flow such as invoking an Airflow DAG, running SQL queries, or otherwise.

An example follows where we'll create a Cloud Build that will run from the Pub/Sub topic triggered once your GA4 BigQuery exports are present. In "Setting Up a Pub/Sub Topic for GA4 BigQuery Exports" on page 114, we created a Pub/Sub topic called "ga4-bigquery," which fires each time the exports are ready. We'll now consume this message via a Cloud Build.

Create a Build Trigger that will respond to the Pub/Sub message. An example is shown in Figure 4-11. For this demonstration, it will read a *cloudbuild.yml* file that is within the `code-examples` GitHub repo. This repo contains the work you wish to do on the BigQuery export for that day.

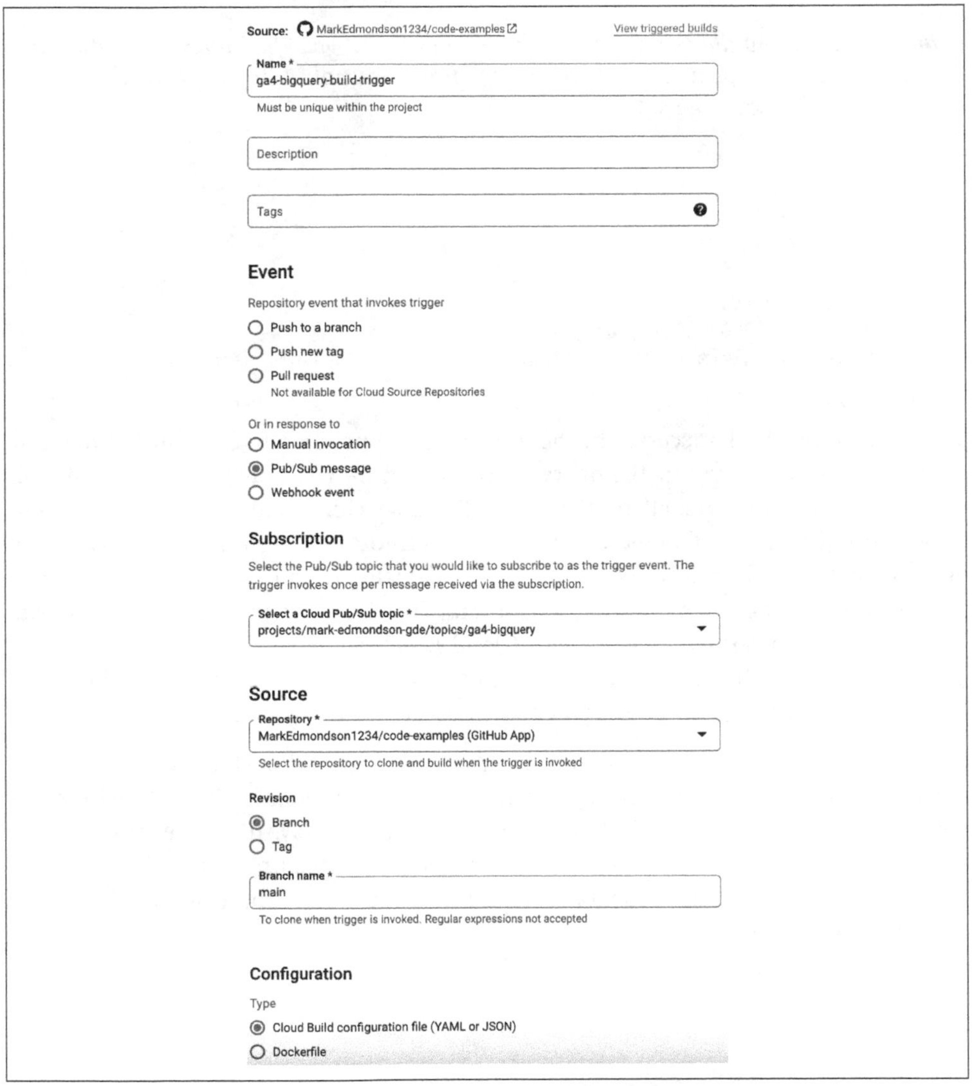

Figure 4-11. Setting up a Build Trigger that will build once the BigQuery export for GA4 is complete

Now we need the build that the Build Trigger will kick off when it gets that Pub/Sub message. We'll adapt the example from Example 4-10 and put this in an SQL file. This is committed to the GitHub source, which the Build will clone before executing. This will allow you to adapt the SQL easily by committing up to GitHub.

Example 4-14. The build the Build Trigger will do when it gets the Pub/Sub event from the GA4 BigQuery export completion; the SQL from Example 4-10 is uploaded in a separate file called ga4-agg.sql

```
steps:
- name: 'gcr.io/cloud-builders/gcloud'
  entrypoint: 'bash'
  dir: 'your/dir/on/Git'
  args: ['-c',
      'bq --location=eu \
      --project_id=$PROJECT_ID query \
      --use_legacy_sql=false \ --destination_table=tidydata.ga4_pageviews \
      < ./ga4-agg.sql']
```

To run Example 4-14 successfully, the user permissions need to be adjusted to allow authorized use to carry out the query. This will not be your own user, since the job will be done on your behalf by the Cloud Build service agent. Within your Cloud Build settings or in the Google Console you will find the service user who will carry out the commands within the Cloud Build, which will look something like 123456789@cloudbuild.gserviceaccount.com. You can use this, or you can create your own custom service account with Cloud Build permissions. This user needs to be added as a BigQuery Admin so they can carry out queries and other BigQuery tasks, such as creating tables, that you may want later. See Figure 4-12.

For your own use cases, you will need to adapt the SQL and perhaps add more steps to work with the data once this step is complete. You can see that Cloud Build is serving a similar role as Cloud Composer but in a simpler way. It's more generic than scheduling queries in BigQuery but not as expensive or feature-rich as Cloud Composer; I find it a nice tool in the toolbox for when you need simple tasks to be scheduled or event-driven.

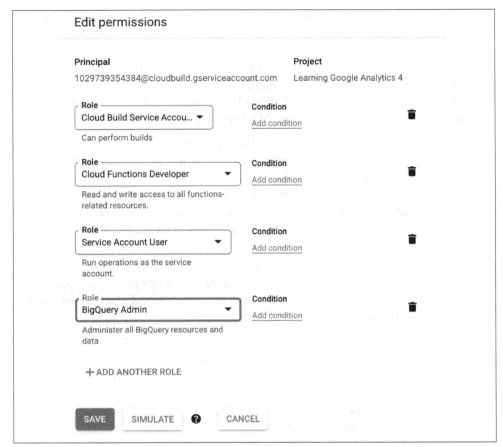

Figure 4-12. Adding the permissions to execute BigQuery jobs to the Cloud Build service account

Cloud Build integrations for CI/CD

Cloud Build can be triggered via schedules, manual invocations, and events. The events cover Pub/Sub and GitHub commits, which are key to its role as a CI/CD tool. In general, it's a good idea when coding to use version control such as Git, and I use GitHub, the most popular version. This way you can keep a record of everything you do and also have an infinite undo to roll back changes if you need to—and when the difference between success and failure can be a . in the wrong place, then this is desirable!

Once you're using Git for version control, you can then also start using it for other purposes, such as triggering builds off of each commit to check your code (tests), make sure it follows style guidelines (linting), or trigger actual builds of the product the code creates.

Cloud Build allows any code language within it via its use of Docker containers, which are used to control the environments of each step. It also features easy authentication for Google Cloud service via `gcloud auth`, as we saw in Figure 4-12, when setting it up to make BigQuery tasks. The `gcloud` commands that you use to deploy services can also be used within Cloud Build to automate those deployments. In addition, all the execution can be based off the code you are committing to the Git repository, so you will have a perfect overview of what is happening and when.

For example, we can deploy DAGs to Airflow, as was done in Example 4-11. Normally, to have your DAG deploy, you need to copy that Python file into a special folder within the Cloud Composer environment, but with Cloud Build you can use `gsutil` (the GCS command-line tool) to do it instead. This encourages more rapid development and gives you some time back to focus on the important things. An example of the cloudbuild file for your trigger is shown in Example 4-15.

Example 4-15. You can deploy python DAGs for Airflow/Cloud Composer using Cloud Build straight from your Git repository—here $\_AIRFLOW\_BUCKET$ is a substitution variable you change to the location of your installation, and the .sql files are assumed to be within a folder named sql *in the same location*

```
steps:
- name: gcr.io/google.com/cloudsdktool/cloud-sdk:alpine
  id: deploy dag
  entrypoint: 'gsutil'
  args: ['mv',
      'dags/ga4-aggregation.py',
      '$_AIRFLOW_BUCKET/dags/ga4-aggregation.py']
- name: gcr.io/google.com/cloudsdktool/cloud-sdk:alpine
  id: remove old SQL
  entrypoint: 'gsutil'
  args: ['rm',
      '-R',
      '${_AIRFLOW_BUCKET}/dags/sql']
- name: gcr.io/google.com/cloudsdktool/cloud-sdk:alpine
  entrypoint: 'gsutil'
  id: add new SQL
  args: ['cp',
      '-R',
      'dags/sql',
      '${_AIRFLOW_BUCKET}/dags/sql']
```

Similar to the previous example for Cloud Composer, Cloud Build can be used to deploy for all other GCP services as well. We use it again in Example 3-15 to deploy Cloud Functions, but any service that uses a gcloud command can be automated.

Batched data scheduling services are usually core to all data applications, including those involving GA4. We've taken a tour through some of your options when looking at scheduling, including BigQuery scheduled queries, Cloud Scheduler, Cloud Build, and Cloud Composer/Airflow. Each have the following advantages and disadvantages:

BigQuery scheduled queries
  Easy to set up but lack accountability and work only for BigQuery

Cloud Scheduler
  Works for all services, but complicated dependencies will start to become hard to maintain

Cloud Build
  Event-based and can trigger from schedules, usually my preferred choice, but does not support flows that need backfills and retries

Cloud Composer
  Comprehensive scheduling tool with backfills, support for complicated workflows, and retry/service level agreement (SLA) features but the most expensive and complicated to work with

Hopefully, this has given you some ideas about what you could use for your own use cases. In the next section, we look at more real-time data flows and the tools you may use when you need to process your data immediately.

Streaming Data Flows

For some workflows, batched scheduling may not be enough. If you're looking for reactive data updates under half an hour, for example, it may be time to start looking at the streaming data flow options. Several of the solutions share some of the same features and components, but there is an increased cost and complexity that comes with real-time streaming data that you need to factor in.

Pub/Sub for Streaming Data

The examples up to now have only treated Pub/Sub with relatively low data volumes, just events to say something has happened. However, its main purpose is for dealing with high-volume data streams, and this is where it really shines. It's at-least-once delivery system means you can build reliable data flows even if you're putting TBs worth of data through it. In fact, Googlebot, the search engine bot that built up Google Search, also runs on a similar infrastructure, and it regularly downloads the entire internet so you know Pub/Sub can scale!

The support for streaming data will most likely start with Pub/Sub as the entry point that other streaming systems send data toward from Kafka or other on-premises systems. When setting up these real-time ingestions, it's usually the internal application developers that will set up this stream before handing it off once they want that stream to flow into the GCP. This is usually where I get involved, with my responsibility being to help define the schema of data coming into the Pub/Sub topic and then taking it onward from there.

Once you have data flowing into a Pub/Sub topic, it has out-of-the-box solutions to start streaming that to popular destinations, such as Cloud Storage and BigQuery. These are provided by Apache Beam or the Google-hosted version, Dataflow.

Apache Beam/DataFlow

The go-to service for streaming data around GCP is Dataflow. Dataflow is a service that runs jobs written in Apache Beam, a data processing library that started off in Google but is now available open source, so you can also use it as a standard for other clouds.

Apache Beam works by creating virtual machines (VMs) with Apache Beam installed that are set up to execute code that will operate on each data packet as it flows in. It has autoscaling built in, so if the machine's resources begin to get stretched (i.e., thresholds on CPU and/or memory are hit), then it will launch another machine and route some of the traffic to that. It will cost more or less depending on how much data you're sending in, with a minimum floor of 1 VM.

There are common data jobs that are expedited for Apache Beam by its templates. For example, a common task is to stream Pub/Sub into BigQuery, which is available without having to write any code at all. An example is shown in Figure 4-13.

To work with the template, you will need to create a bucket and the BigQuery table the Pub/Sub messages will flow into. The BigQuery table needs to have the correct schema that should match the Pub/Sub data schema.

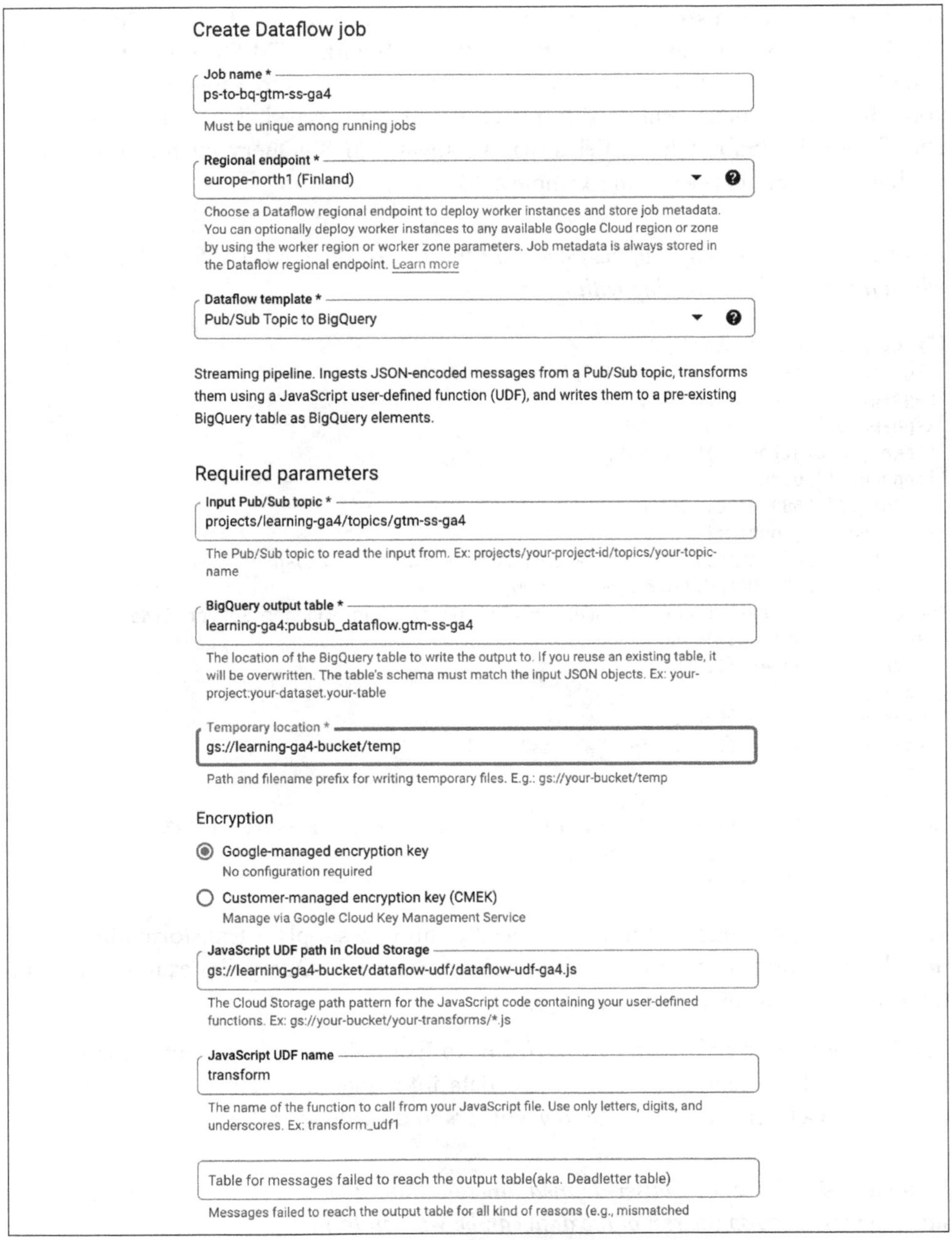

Create Dataflow job

Job name *
ps-to-bq-gtm-ss-ga4

Must be unique among running jobs

Regional endpoint *   ▼   ❷
europe-north1 (Finland)

Choose a Dataflow regional endpoint to deploy worker instances and store job metadata.
You can optionally deploy worker instances to any available Google Cloud region or zone
by using the worker region or worker zone parameters. Job metadata is always stored in
the Dataflow regional endpoint. Learn more

Dataflow template *   ▼   ❷
Pub/Sub Topic to BigQuery

Streaming pipeline. Ingests JSON-encoded messages from a Pub/Sub topic, transforms
them using a JavaScript user-defined function (UDF), and writes them to a pre-existing
BigQuery table as BigQuery elements.

Required parameters

Input Pub/Sub topic *
projects/learning-ga4/topics/gtm-ss-ga4

The Pub/Sub topic to read the input from. Ex: projects/your-project-id/topics/your-topic-
name

BigQuery output table *
learning-ga4:pubsub_dataflow.gtm-ss-ga4

The location of the BigQuery table to write the output to. If you reuse an existing table, it
will be overwritten. The table's schema must match the input JSON objects. Ex: your-
project:your-dataset.your-table

Temporary location *
gs://learning-ga4-bucket/temp

Path and filename prefix for writing temporary files. E.g.: gs://your-bucket/temp

Encryption

◉ Google-managed encryption key
    No configuration required

◯ Customer-managed encryption key (CMEK)
    Manage via Google Cloud Key Management Service

JavaScript UDF path in Cloud Storage
gs://learning-ga4-bucket/dataflow-udf/dataflow-udf-ga4.js

The Cloud Storage path pattern for the JavaScript code containing your user-defined
functions. Ex: gs://your-bucket/your-transforms/*.js

JavaScript UDF name
transform

The name of the function to call from your JavaScript file. Use only letters, digits, and
underscores. Ex: transform_udf1

Table for messages failed to reach the output table(aka. Deadletter table)

Messages failed to reach the output table for all kind of reasons (e.g., mismatched

*Figure 4-13. Setting up a Dataflow from within the Google Cloud Console for a Pub/Sub
topic into BigQuery via the predefined template*

For my example, I'm streaming in some GA4 events from my blog into Pub/Sub via GTM SS (see "Streaming GA4 events into Pub/Sub with GTM SS" on page 227). By default, the stream will attempt to write every Pub/Sub field to a BigQuery table, and your BigQuery schema will need to match exactly to succeed. This can be problematic if your Pub/Sub includes fields that are invalid in BigQuery, such as those with hyphens (-) that are present in Example 4-16.

Example 4-16. An example of the JSON sent from a GA4 tag in GTM SS to Pub/Sub, which has some fields starting with x-ga

```
{"x-ga-protocol_version":"2",
"x-ga-measurement_id":"G-43MXXXX",
"x-ga-gtm_version":"2reba1",
"x-ga-page_id":1015778133,
"screen_resolution":"1536x864",
"language":"ru-ru",
"client_id":"68920138.12345678",
"x-ga-request_count":1,
"page_location":"https://code.markedmondson.me/data-privacy-gtm/",
"page_referrer":"https://www.google.com/",
"page_title":"Data Privacy Engineering with Google Tag Manager Server Side and ...",
"ga_session_id":"12343456",
"ga_session_number":1,
"x-ga-mp2-seg":"0",
"event_name":"page_view",
"x-ga-system_properties":{"fv":"2","ss":"1"},
"debug_mode":"true",
"ip_override":"78.140.192.76",
"user_agent":"Mozilla/5.0 (Windows NT 10.0; Win64; x64) AppleWebKit/537.36 ...",
"x-ga-gcs-origin":"not-specified",
"user_id":"123445678"}
```

To accommodate your customization needs, you can supply a transformation function that will modify the stream before it passes it into BigQuery. For example, we can filter out the fields that start with x-ga.

The Dataflow user-defined function (UDF) in Example 4-17 filters out those events, so the rest of the template can send the data into BigQuery. This UDF needs to be uploaded to a bucket for the Dataflow workers to download and use it.

Example 4-17. A Dataflow user-defined function that filters out Pub/Sub topic fields starting with x-ga so the rest of the data can be written to BigQuery

```
/**
 * A transform function that filters out fields starting with x-ga
 * @param {string} inJSON
 * @return {string} outJSON
 */
```

```
function transform(inJSON) {
 var obj = JSON.parse(inJSON);
 var keys = Object.keys(obj);
 var outJSON = {};

 // don't output keys that starts with x-ga
 var outJSON = keys.filter(function(key) {
   return !key.startsWith('x-ga');
 }).reduce(function(acc, key) {
   acc[key] = obj[key];
   return acc;
 }, {});

 return JSON.stringify(outJSON);
}
```

Once the Dataflow job is set up, it will give you a DAG much like Cloud Composer/ Airflow, but in this system it will be dealing with real-time event-based flows, not batches. Figure 4-14 shows what you should see in your Dataflow Jobs section in the web console.

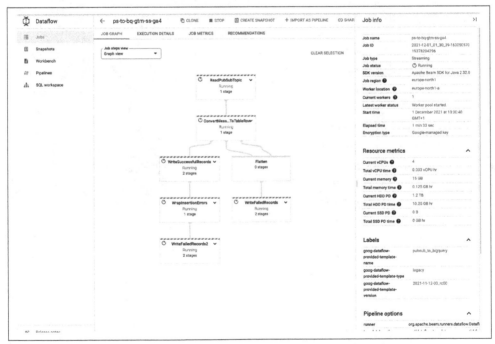

Figure 4-14. Starting up a running job for importing Pub/Sub messages into BigQuery in real time

Costs of Dataflow

Given the way Dataflow works, be wary of spinning up too many VMs because if you get an error, that can send a lot of hits into your pipeline, and you can quickly run up expensive bills. Once you have an idea of your workload, it's wise to set an upper limit on the VMs to deal with the spikes of data but not let it run away from you if something really unexpected happens. Even with these precautions, the solution is still more expensive than batched workflows, since you can expect to spend in the region of $10 to $30 a day or $300 to $900 a month.

The BigQuery schema needs to match the Pub/Sub configuration, so we need to create a table. The table in Figure 4-15 is also set up to be time partitioned.

gtm-ss-ga4

Field name	Type	Mode	Policy tags ⓘ
event_name	STRING	NULLABLE	
engagement_time_msec	INTEGER	NULLABLE	
debug_mode	STRING	NULLABLE	
author	STRING	NULLABLE	
category	STRING	NULLABLE	
published	STRING	NULLABLE	
words	STRING	NULLABLE	
read_time	STRING	NULLABLE	
screen_resolution	STRING	NULLABLE	
language	STRING	NULLABLE	
client_id	STRING	NULLABLE	
page_location	STRING	NULLABLE	
page_referrer	STRING	NULLABLE	
page_title	STRING	NULLABLE	
ga_session_id	STRING	NULLABLE	
ga_session_number	INTEGER	NULLABLE	
user_id	STRING	NULLABLE	
ip_override	STRING	NULLABLE	
user_agent	STRING	NULLABLE	

Figure 4-15. The BigQuery data schema to receive the Pub/Sub JSON

If you make any mistakes, the Dataflow job will stream the raw data into another table in the same dataset where you can examine the errors and make corrections, as shown in Figure 4-16.

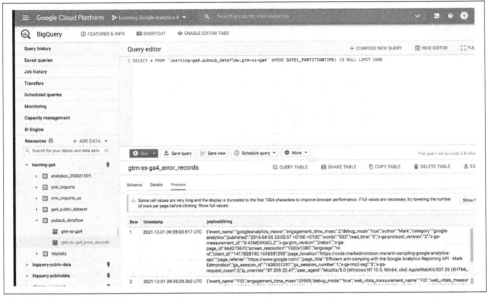

Figure 4-16. Any errors from the Dataflow will appear in its own BigQuery table so you can examine the payloads

If everything is going well, you should see your Pub/Sub data start to appear in BigQuery—give yourself a pat on the back if you see something similar to Figure 4-17.

A standard BigQuery export functionality is already available to you for free by using GA4's native BigQuery exports, but this process can be adapted for other use cases directing to different endpoints or making different transformations. An example may be to work with a subset of your GA4 events to modify the hit to be more privacy aware or to enrich it with product metadata available only via another real-time stream.

Remember that Dataflow will be running a VM at an expense for this flow, so turn it off if you don't need it. If your data volumes aren't large enough to warrant such an expense, you can also use Cloud Functions to stream data.

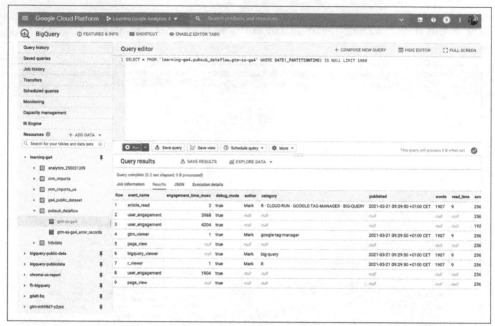

Figure 4-17. A successful streaming import from GA4 into GTM SS to Pub/Sub to BigQuery

Streaming Via Cloud Functions

If your data volumes are within Cloud Function quotas, your Pub/Sub topic setup can also use Cloud Functions to stream the data to different locations. Example 4-5 has some example code for sporadic events like BigQuery tables, but you can also react to more regular streams of data and Cloud Function will scale up and down as needed—each invocation of the Pub/Sub event will create a Cloud Function instance that will run in parallel with other functions.

Limits (for Generation 1 Cloud Functions) include only 540 seconds (9 minutes) of runtime and a total of 3,000 seconds of simultaneous invocations (e.g., if a function takes 100 seconds to execute, you can have up to 30 functions running at a time). This means you should make your Cloud Functions small and efficient.

The following Cloud Function should be small enough that you can have around 300 requests per second (Example 4-18). It takes the Pub/Sub message and puts it as a string into a raw data column in BigQuery along with its timestamp. You can modify the code to parse out more specific schema as you need it or use BigQuery SQL itself to process the raw JSON string into tidier data later.

Example 4-18. Modify the pb dict within the code to parse out more fields if you want to create a more bespoke table. Add environment arguments `dataset` and `table` pointing at your premade BigQuery table. Inspired by Milosevic's Medium post (https://oreil.ly/ Zuy4u) about how to copy data from Pub/Sub to BigQuery.

```python
# python 3.7
# pip google-cloud-bigquery==2.23.2
from google.cloud import bigquery
import base64, JSON, sys, os, time

def Pub/Sub_to_bigq(event, context):
  Pub/Sub_message = base64.b64decode(event['data']).decode('utf-8')
  print(Pub/Sub_message)
  pb = JSON.loads(Pub/Sub_message)
  raw = JSON.dumps(pb)

  pb['timestamp'] = time.time()
  pb['raw'] = raw
  to_bigquery(os.getenv['dataset'], os.getenv['table'], pb)

def to_bigquery(dataset, table, document):
  bigquery_client = bigquery.Client()
  dataset_ref = bigquery_client.dataset(dataset)
  table_ref = dataset_ref.table(table)
  table = bigquery_client.get_table(table_ref)
  errors = bigquery_client.insert_rows(table, [document], ignore_unknown_values=True)
  if errors != [] :
    print(errors, file=sys.stderr)
```

The function takes environment arguments to specify where the data goes, as shown in Figure 4-18. This enables you to deploy multiple functions for different streams.

The premade BigQuery table has only two fields, raw, which contains the JSON string, and timestamp, when the Cloud Function executed. You can use BigQuery's JSON functions (*https://oreil.ly/AXxOL*) in SQL to parse out this raw JSON string, as shown in Example 4-19.

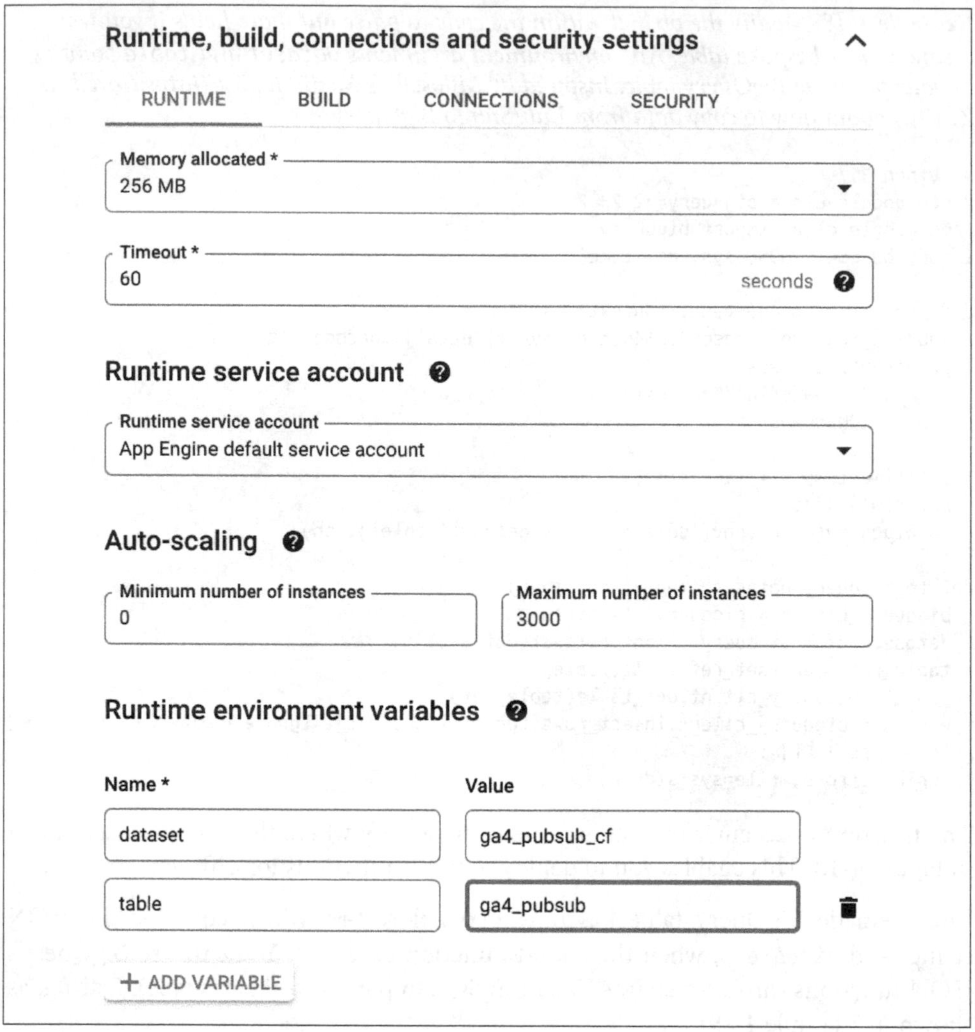

Figure 4-18. Setting the environment arguments for use within the Cloud Function in Example 4-18

Example 4-19. BigQuery SQL to parse out a raw JSON string

```
SELECT
 JSON_VALUE(raw, "$.event_name") AS event_name,
 JSON_VALUE(raw, "$.client_id") AS client_id,
 JSON_VALUE(raw, "$.page_location") AS page_location,
 timestamp,
 raw
FROM
 `learning-ga4.ga4_Pub/Sub_cf.ga4_Pub/Sub`
WHERE
```

```
DATE(_PARTITIONTIME) IS NULL
LIMIT
 1000
```

The result of the code in Example 4-19 is shown in Figure 4-19. You can set up this SQL downstream in a schedule or via a BigQuery View.

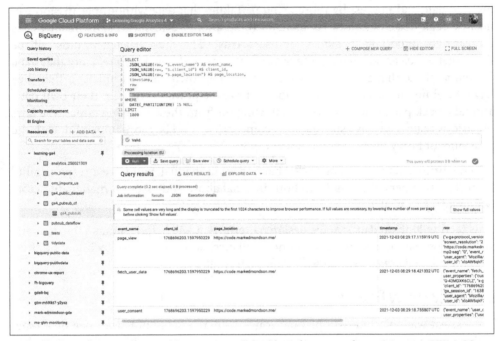

Figure 4-19. The raw data table receiving the Pub/Sub stream from GA4 via GTM-SS can have its JSON parsed out with BigQuery's functions such as `JSON_VALUE()`

Streaming data services offer a way to have the most responsive and modern data stack for your GA4 set up but they do come at a financial and technological cost that you will need to justify with a great business case. However, if you have that case, having these tools available means you should be able to get something up and running, an application that would have been almost impossible to do 10 years ago.

Digital analytics data streams such as GA4 are usually the most useful when they can tailor experiences to an individual user, but when using data that can be associated with a person, you need to be extra mindful of the consequences both legally and ethically. We'll delve into this in the next section.

Protecting User Privacy

"User Privacy" on page 40 also gives an overview of user privacy, but this section goes into more depth and gives some technical resources.

In this modern data age, the value of data has been realized by those who are providing that data and not just by those using it. Taking that data without permission can now be regarded as immoral as theft, so to be a long-term sustainable business, it's increasingly important to gain your users' trust. The most trusted brands will be those that are clear on what data they are capturing and how they are using that data, as well as those that give users easy access to their own data, so much so that they have the ability to make informed choices, have the power to reclaim their own data, and take back permission to use it. Following on from these evolving ethics, the laws of various regions are also starting to become stricter and be more impactful, with the possibilty of heavy fines if you do not comply.

When storing data that can be traced back to an individual, you have a responsibility to protect that personal data from both internal abuse and external malignant actors who may try to steal it.

This section looks at data storage design patterns that will help you make the data privacy process easier. In some cases, noncompliance has occurred not by intent but by a poorly designed system, which we'll look to prevent.

Data Privacy by Design

The easiest way to avoid data privacy concerns is to simply not store personal data. Unless you have a specific need for that personal data, removing it from your data capture or erasing it as it comes into storage is the easiest way to keep on top of it. This may sound flippant, but it does need to be stated, since it's quite common for companies to just collect this data by accident or without really thinking about the consequences. A classic case with web analytics is accidentally storing user emails in the URLs for web forms or in search boxes. Even if accidental, this is against Google Analytics Terms of Service and carries the risk of having your account closed down. Having some data cleaning in place at the point of collection can go a long way toward keeping a clean house.

If you do need some level of personalization, you still need not necessarily capture data that will pose a privacy risk. This is where pseudonomyzation comes in, which is the default for data collection, including GA4. Here's an ID is assigned to a user, but it is that ID that is shared, not the user's personal data. An example is choosing between a random ID or a person's phone number as a user ID. If the random ID is accidentally exposed, an attacker couldn't do much with it unless they had access to the system that mapped that ID to the rest of that person's information. If the ID exposed is actually a person's phone number, the attacker has something they can use immediately. Using pseudonomyzation is a first line of defense for guarding a user's personal data. Again, never use email or a phone number as an ID because you will run into privacy breaches; companies who have done this have been fined.

It may be that the ID is all you need for your use cases—for instance, the default GA4 data collection is at this level. Only when you start to link that ID to personal information, such as linking GA4's `client_id` to a user's email address, will you need to start considering more extreme privacy considerations. This typically happens when you start linking it to your backend systems like a CRM database.

Should your use cases then require personal data, email, name, or otherwise, there are some principles to keep in place, encouraged by privacy legislation such as GDPR. These steps will allow you to preserve the user's dignity as well as have some business impact with that data:

Keep personal data (PII) in minimal number of locations
Personal data should be kept in as few databases as possible and then joined or linked to other systems via a pseudonymous ID that is the key for that table. This way you should only ever have one place to look if you need to delete or extract a person's data, and you won't need to also delete it from subsequent places it may have been copied or joined to. This complements the next point about encryption of user data, since you should only need to do that with the user database.

Encrypt user data with salt-and-peppered hashes
The process of hashing is a method of one-way encrypting data so that it is impossible to re-create the original data without knowing the ingredients: for instance, "Mark Edmondson" when hashed with the popular sha256 hashing algorithm is:

```
3e7e793f2b41a8f9c703898c5c0d4e08ab2f22aa1603f8d0f6e4872a8f542335
```

However, it is always this hash and should be globally unique so you can use it as a reliable key. To "salt and pepper" the hash means you also add a unique keyword to the data to make it even more secure should somehow someone break the hashing algorithm or can obtain the same hash to make a link. For instance, if my salt is "baboons," then I prefix it to my data point so that "Mark Edmondson" becomes "baboonsMark Edmondson," and the hash is:

```
a776b81a2a6b1c2fc787ea0a21932047b080b1f08e7bc6d6a2ccd1fb6443df48
```

e.g., completely different than before. Salts can be global or stored with the user point to make them unique for each user. To "pepper" or "secret salt" the hash is a similar concept, but this time the keyword is not kept alongside the data to encrypt but in another secure location. This guards against database breaches because it now has two locations. In this case, the "pepper" would be fetched and may be "averylongSECRETthatnoonecanknow?" and so my final hash will be

"baboonsMark EdmondsonaverylongSECRETthatnoonecanknow?" to give a
final hash of:

```
c9299fe251319ffa7ec66137acfe81c75ee115ceaa89b3e74b521a0b5e12d138
```

which should be very difficult for a motivated hacker to reidentify the user with.

Put data expiration times on personal data

Sometimes you will have no option but to copy personal data across, for instance,
if you're importing from different clouds or systems. In that case, you can nomi-
nate the data source as the source of truth for all your privacy initiatives and then
enforce a data expiration date on any data that is copied from that source. Thirty
days is typical, meaning you need to do a full import at least every 30 days
(maybe even daily) so the data volumes will increase. You should then be secure
knowing that as you update user permissions and values in your master database,
any copies of that data will be ephemeral and not be around at all once the
imports stop.

Adding privacy principles does add extra work, but the payoff is peace of mind and
trust in your own systems, which can be conveyed to your customers. An example of
the last point for data expiration within some of the storage systems we've talked
about in this chapter is demonstrated in the next section.

Data Expiration in BigQuery

When setting up your datasets, tables, and buckets, you can set the data expiration for
your data that is coming in. We've already covered how to set it in GCS in "Google
Cloud Storage" on page 82.

For BigQuery, you can set an expiration date at the dataset level that will affect all
tables within that dataset—see Figure 4-20 for an example with a test dataset.

Dataset info ✏	
Dataset ID	learning-ga4:tests
Created	2 Dec 2021, 10:15:24
Default table expiry	30 days 0 hr
Last modified	2 Dec 2021, 10:15:24
Data location	EU

Figure 4-20. You can configure the table expiration time when you create a dataset

For partitioned tables, you will need something different, since the table will always exist but you want the partitions themselves to expire over time, leaving you with only the most recent data. For that, you will need to invoke gcloud or use the Big-Query SQL for altering table properties, as shown in Example 4-20.

Example 4-20. Setting a date expiration for BigQuery partitions in a partition table

The gcloud way (via your local bash console or in Cloud Console):

```
bq update --time_partitioning_field=event_date \
  --time_partitioning_expiration 604800 [PROJECT-ID]:[DATASET].partitioned_table
```

Or via BigQuery DML:

```
ALTER TABLE `project-name`.dataset_name.table_name
SET OPTIONS (partition_expiration_days=7);
```

As well as the passive data expiration times, you can also actively scan data for privacy breaches via the Data Loss Prevention API.

Data Loss Prevention API

The Data Loss Prevention (DLP) API is a way to automatically detect and mask sensitive data such as emails, phone numbers, and credit card numbers. You can call it and run it on your data within Cloud Storage or BigQuery.

If you have a large amount of streaming data, there is a Dataflow template available to read CSV data from GCS and put the redacted data into BigQuery (*https://oreil.ly/dFlye*).

For GA4, you can most easily use it to scan your BigQuery exports to see if any personal data has been collected inadvertently. The DLP API scans only one table at a time, so the best way to use this is to scan your incoming data table each day. If you have a lot of data, I recommended scanning only a sample and/or restrict the scan to only the fields that may contain sensitive data. For GA4 BigQuery exports in particular, this is most likely to be only the event_params.value.string_value since all other fields are more or less fixed by your configuration (event_name, etc.).

Summary

Because data comes in such a variety of forms and uses, there are many different systems available to hold it. The broad categories we have spoken about in this chapter are structured and unstructured data and scheduled versus streaming pipelines between those systems. You also need a good organizing structure with some thought on who and what should access each piece of data along its journey, since at the end of the day, it's people who will be using that data, and lowering the friction for the right person to see the right data is a big step toward data maturity within your organization. As soon as you need data beyond GA4, you need to know how these systems interact, but you have a good starting point with the GA4 BigQuery exports, which is highly regarded as one of its key features over Universal Analytics.

Now that we've talked about how to collect and store data, in the next chapter, we'll get more into how that data is actively massaged, transformed, and modeled, and how it usually represents where the most value is added within the pipeline.

## CHAPTER 5

Data Modeling

Data modeling has the potential to be the most technical aspect of a project, since this is typically where you would see machine learning or advanced statistics working across your data. It could also be as simple as making a join across two datasets. It is here where the magic happens in a data project, turning that raw data into information and from there into insight. However, data modeling shouldn't be the end goal—that is reserved for when we export that insight to the data activation channels, be it analysis in a one-off report or pushing to your data activation channels. Data modeling is a means to an end, not the end to the means. It should be about how you extract value out of your data and never about using the latest technique—it may be that the task that will extract the best value is a simple join rather than a sophisticated neural net. I also think of data modeling as the place where you put your own unique business logic that defines your advantages over your competition. Here's where you can be creative and bring your own competitive advantage and experience, tailoring how your data is used for the eventual end goal of helping your customers and your business.

There are many ways to model your data outside of GA4, which we will dig into, but we first turn to what GA4 provides natively within its own platform.

GA4 Data Modeling

When using GA4, you can take advantage of some data modeling that comes baked into the product that saves you from needing to create or customize your own. This will be an ever-evolving feature set of GA4, since one of the reasons GA4 has a new event-based data schema is to enable these applications more easily.

At the time of writing, there are several data modeling choices available within GA4.

Standard Reports and Explorations

The raw data flowing into GA4 isn't useful day-to-day, as evidenced if you try to work with the BigQuery exports directly. Most people don't want to have to type out some SQL before they see some information about how many users came from each channel, for example. This is where the pre-created default reports and the customizable explorations sections come into play. GA4 uses Report Libraries to define what reports each user will see when they log in, which can be configured toward the "less is more" mindset of not overloading end users with every report. Hopefully, these will be expanded to include typical use case scenarios such as ecommerce, publisher, blog analytics, etc. Using the transformations for those reports is talked about in "Visualization" on page 203.

Attribution Modeling

How you assign credits for conversions within any analytics system is subject to your attribution model. Even if you're not actively looking at attribution reports, you are implicitly choosing the default settings for your analytics system. This applied to Universal Analytics even when you weren't looking at the dedicated multitouch or attribution reports—the rest of the reports were using a "last nondirect channel" model (see "GA4 Attribution Choices" on page 163).

In GA4, you have more configuration options for how to attribute conversions throughout your account. Within your GA4 configurations for Attribution Settings, you can select options such as "Cross channel last click" or "Ads-preferred last click" as well as the look-back window on how long a channel should be credited for a future conversion; see Figure 5-1.

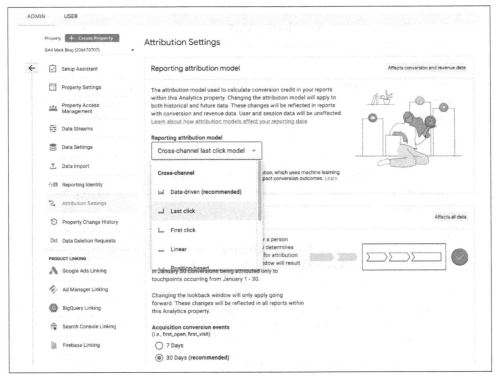

Figure 5-1. You can set attribution settings for how your conversions are attributed to channels and select the lookback window for 30, 60, or 90 days

GA4 Attribution Choices

There are several ways to attribute conversions in GA4 that work throughout the reports. This setting can be changed at any time and applies retroactively to historical data. Here is a brief overview:

Data driven

This uses its own machine learning algorithms to find a model that learns how different GA4 events impact your conversions. It means those events that did and did not contribute to a conversion, and compares what would have happened if that event was not present to build up a model of what weighting that event should have for new conversions.

Last nondirect click

This attributes conversions to the last click but only if it's a nondirect click, e.g., a journey of organic → paid → direct will attribute to paid. This is what Universal Analytics used for its standard reports. (Recall Direct is any visit that cannot be attributed to a channel source, such as untagged campaigns, bookmarks, or direct navigation).

First click

> This attributes conversions to the first click, where the user was first seen. A journey of organic → paid → direct will attribute to organic.

Linear

> This registers a fraction of each goal to all the channels that contributed to the conversion. A journey of organic → paid will attribute 50% to organic and 50% to paid.

Position-based

> This attributes 40% credit to the first and last interactions and then uses linear attribution to allocate the remaining 20% in the middle. A journey of organic → paid → organic → email will attribute 40% + 10% to organic, 40% to email, and 10% to paid.

Time-decay

> This uses a 7-day half-life decay to attribute conversions. A click 8 days before a conversion gets 50% less credit than a click 1 day before.

Ads-preferred

> This ignores all channels that are not Google Ads and attributes 100% of value to that, unless no Google Ads click is present, in which case it will fall back to the last nondirect click.

Considering your GA4 usage carefully should influence this decision: an ecommerce web shop where most users buy on their first or second session will have different needs than a luxury car–brand website whose customers may be considering purchases for many months.

User and Session Resolution

Resolving which hits belong to a user is a nontrivial exercise, especially in these modern times when cookie lifetimes are not reliable. Within your GA4 settings, there is a "Reporting Identity" setting where you can decide how GA4 will model users and their sessions—either by user ID plus device ID or by device ID only (see Figure 5-2).

This can potentially heavily change your user metrics if you're also sending in a `userId` to identify users or if you're using Google Signals. Refer to "Google Signals" on page 68 for more details about these settings. Assuming you have consent from those users, a user ID will more reliably identify them even if they arrive via desktop versus mobile or with different cookies.

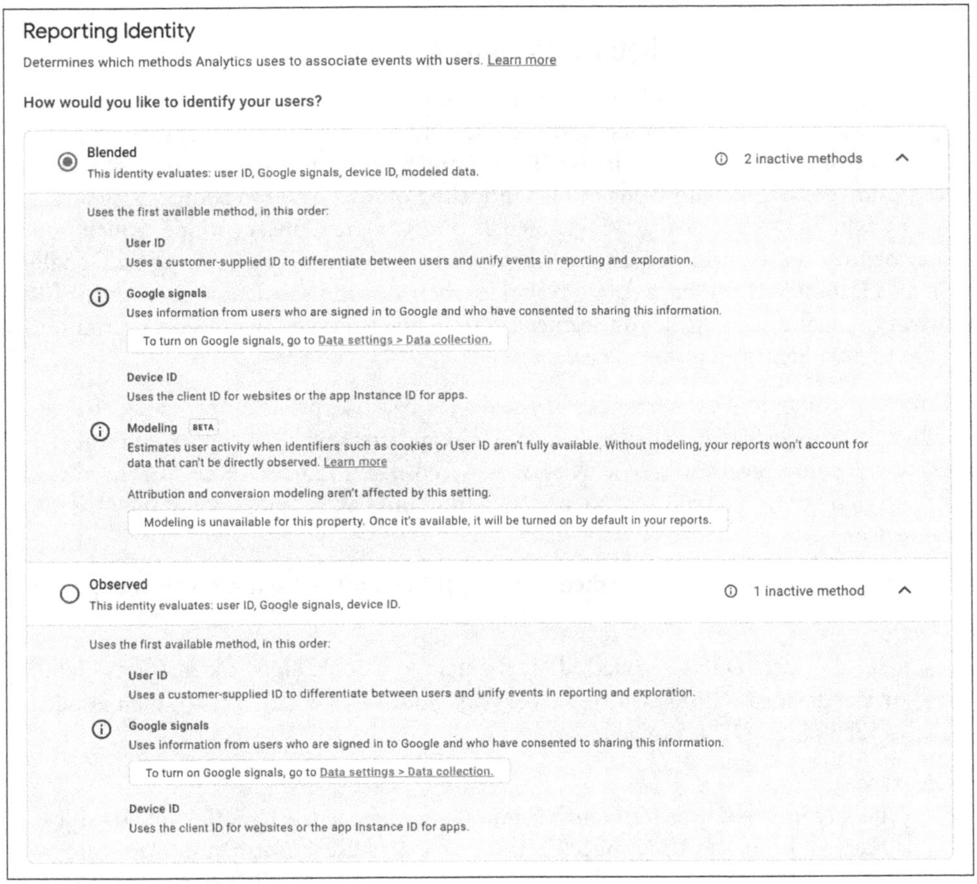

Figure 5-2. Selecting how users can be identified within GA4 reports

Consent Mode Modeling

Google Consent Mode (*https://oreil.ly/D0aK7*) is a new technology that hooks into cookie management systems to help respect user choices about whether cookies should be dropped or not. If they choose not to, a cookieId (cid) cannot be assigned to that user, so each GA hit being collected will be generated with a new cid. If these hits were allowed to be surfaced in the GA4 reports by default, then user metrics would be much inflated. As a result, these hits are not surfaced in your GA4 account. However, by comparing the traffic patterns of users who do consent with those who do not, GA4 can give you an idea of what attribution would be in place had everyone consented to cookies, e.g., if 50% of your consented users are attributed to paid search, there is a good case that 50% of your nonconsented users will also be attributed to paid search.

Reporting Identity Spaces

The user and consent mode resolution is codified in GA4's configurable identity spaces, which let you determine how users should be assigned for your own website. Some websites that always use a user ID may prefer to use this method only so as not to muddy the waters with Google's own modeling on top. The modeling is also affected by whether you're using Google Signals and Google Consent Mode, which you may or may not be using depending on your user privacy policy. Turning on Google Signals also means you have data thresholds when viewing the data to avoid identifying individual users (e.g., if you segment to only one user, you will need to zoom out so as to only identify a group of customers).

Universal Analytics's default would be the device-based approach, only using cookies with cid. Google Signals relies on logged-in Google users who have given consent at a Google account level to share activity across websites. Modeling users will need the stream of unconsented hits that have no identifier but can be guessed at if they are the same user.

With those approaches, you can then decide within your GA4 settings which to pick:

Blended
> Uses the user ID if it is collected, then defers to Google Signals if that is available or the device ID if that isn't available (e.g., your GA4 cookie [cid]), then applies modeling if no ID is available.

Observed
> Chooses between user ID, Google Signals, then device ID. Uses the user ID if it is collected. If no user ID is collected, then Analytics uses information from Google Signals if that is available. If neither user ID nor Google Signals information is available, then Analytics uses the device ID.

Device-based
> Uses only the device ID and ignores any other IDs that are collected.

Audience Creation

Audiences in GA4 replace the concept of segments in Universal Analytics and are a subset of your traffic created by you or based on what GA4 defines. Unlike Universal Analytics segments, however, Audiences are not retroactive for historical traffic, so it's important to define them sooner rather than later if you intend to use them later for data activation. The rules for creating these segments can come from the custom dimensions and events you have configured for your GA4 data collection, and indeed Audiences may be the reason you configure particular custom fields in the first place. GA4's Audiences are a route to activating data within the entire GMP and therefore are usually the first place to look for your data activation channels. See "GA4 Audiences and Google Marketing Platform" on page 195 for some examples.

Predictive Metrics

Out of the box, GA4 comes with three predictive metrics: purchase, churn, and revenue predictions. This generates a prediction for every user who visits your website. Predictive metrics appear only if your website fulfills certain prerequisites such as traffic volume. Being able to use predictions within your data represents a major step-up in capabilities for GA4 over Universal Analytics. Once you have a predictive metric, you can then couple it with the GA4 Audiences to export those users who are predicted to transact or churn so you can act on that information. This use case is examined in detail in Chapter 7.

Insights

Insights is a GA4 feature that constantly runs machine learning models such as anomaly detection to help surface information from your data that may have been hidden from view. It combines this with a natural language processor interface so you can write, "What was my lowest converting channel?" in the search bar at the top of the page and it will look to surface the appropriate report for you—see Figure 5-3 for an example.

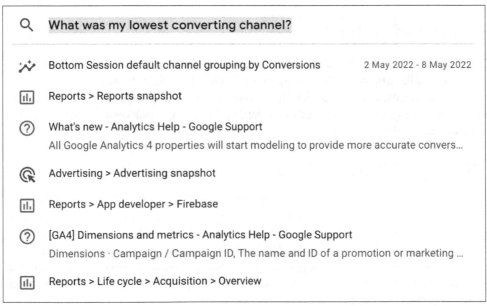

Figure 5-3. Writing questions in the GA4 search bar will parse itself to try and find the most appropriate GA4 report for you

Insights will try to flag the most important findings it has and display them on the GA4 home page, and you can also access them directly in the Insights sidebar. The example in Figure 5-4 shows some findings of forecast versus actual trends, top performers, and spikes in metrics. Insights may help if you find a serendipitous finding that aligns with your aims that day, but in general you will need more concrete goals and aims to derive value out of it via its natural language parsing to give you a Q&A interface to explore your data rather than trying to construct your own reports. It lowers amount of effort required for users looking for information quickly within GA4's interface.

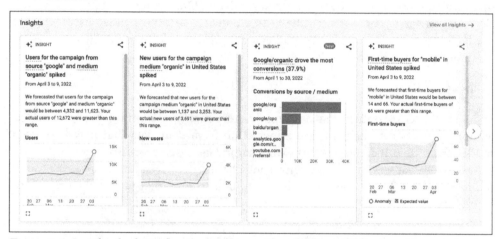

Figure 5-4. Insights looks to flag the most important findings of the day when you log in

The GA4 modeling features have been chosen to cover some of the most common use cases digital analytics users need to work with their data; however it's impossible to cover all needs. Should you identify a use case beyond the defaults, you'll need to start considering extracting your GA4 data and creating your own modeling, which we'll discuss in the next section.

Turning Data into Insight

Consider your business goals as what lies underneath the big "X" on a treasure map. Different map locations are where you're storing your data, but it is your data modeling that will join the dots and guide you along your journey. If you have been following the process outlined in Chapter 2, you should by now have scoped out a business objective and have all the data you need in storage. But now you're looking to turn that data into something useful. Data modeling is responsible for that transformation, so we now turn to how to shape that process.

We'll cover attributes of your data and consider what your data modeling will need to uncover: drilling down into the precise metrics you will need to measure performance and error rates, seeing how different data types will call for different modeling techniques, and considering some common scenarios such as linking datasets as a method to gain your insights.

Scoping Data Outcomes

When considering your modeling, the more accurately you can define what the output will be, the clearer route you will have to achieve it. This drills down how many outcomes you need, the type of data it represents (numerical, categorical, etc.), the exact definition of that data, and what the data is recording. From a technical data modeling perspective, certain categories of data will work better or worse with certain modeling techniques. We'll go over some of these here:

Accuracy versus precision versus recall

Naively, we may always want our models to be as accurate as possible, but that is not a precise-enough word to enable all modeling performance. When scoping out the key performance metric for your models, this can make a huge difference. For instance, say you have a conversion rate of 1% on your website, and you wish to make a model to predict a user's conversion rate. If you plug in accuracy without thinking, you may end up with a model that is 99% accurate because it predicts that all users will not convert. Strictly true, but unhelpful! A better metric is needed, which is likely precision or recall.

Type I Versus Type II Errors

A key concept here is that unless you have 100% accuracy (never happens), you will get two different types of errors: false positives that say a user converted but they didn't and false negatives that say a user didn't convert when they did. These are defined as *type I errors* (you incorrectly predict a conversion when they did not) and *type II errors* (you incorrectly predict a nonconversion when they did. See Table 5-1.

My favorite mnemonic for this is a doctor talking to a man and a pregnant woman: "You're pregnant!" the doctor tells the man (type I); "You're not pregnant!" the doctor tells the (pregnant) woman (type II). As you can imagine for this case, it's a lot more important to lower type II errors than minimize type I, but perhaps some new glasses will help too.

The exact definitions for accuracy, precision, and recall are shown in Example 5-1.

Example 5-1. Defining accuracy versus precision versus recall

```
Accuracy = (True Positive + True Negative) / All Results

Precision = True Positive / (True Positive + False Positive)

Recall = True Positive / (True Positive + False Negative)
```

Take the preceding example about trying to predict a 1% conversion rate and our model returns that no one converts. In that case, supposing we have 1,000 total visits with 100 conversions:

- Accuracy = (0 + 990) / 1000 = 99%
- Precision = 0
- Recall = 0

An accurate model but not a helpful one.

We can plot model results in a confusion matrix as shown in Table 5-1 to help decide what measure we should use.

Table 5-1. Confusion matrix

	Prediction TRUE	Prediction FALSE
Actual TRUE	Correct! (true positive)	Type II error (false negative)
Actual FALSE	Type I error (false positive)	Correct! (true negative)

As a more realistic example, let's fill in some numbers and suppose our model creates a confusion matrix like that shown in Table 5-2.

Table 5-2. Confusion matrix for a more realistic model

	Prediction TRUE	Prediction FALSE
Actual TRUE	9	1
Actual FALSE	90	900

Here we can see that the model caught most of the real conversions but also predicted that a lot of other users would convert who did not. Plugging in the numbers to the formula shown in Example 5-1, we get the following:

- Accuracy = (9 + 900) / 1000 = 90.9%
- Precision = 9 / (9 + 90) = 0.09
- Recall = 9 / (9 + 900) = 0.9

Note that the accuracy has gone down compared to our previous base "no one converts" model, but our recall and precision are now nonzero values and are preferable as metrics.

The exact metric you use will depend on what is most important for your use case. If you want to make sure you include the most correct results and don't mind if some people are mislabeled as correct, then recall may be your metric. If you want to make sure you minimize the number of people you label as converts, then precision may be preferable. If you would like a balance between the two, you can use the F_1-score (*https://oreil.ly/lafOP*).

Classification versus regression
> The main difference here is akin to working with dimensions or metrics. With classification, you'll want to, say, predict which channel a user is arriving by, which could be one of 8 values, for example. With regression, you will be working with continuous variables such as number of page views. Middle ground between these two is logistic regression, which is a prediction of a 0 or a 1 or TRUE versus FALSE: this could be if a user transacts or not, for example. The main reason you may want to define this at the start of the project is that it will largely depend on which statistical or machine learning technique you use. Things can go awry if you accidentally use a classification model for a regression problem, for example.

Rates versus counts versus ranks
> Another important type of data is the type of thing it is measuring, since like classification and regression, it can heavily impact what models are available. Rates such as conversion rates and bounce rates are typically standardized to run from 0% to 100%, whereas counts can be any real number both positive or negative. Ranks such as search position typically start at 1 and are strictly positive. Realizing there is a difference to these numbers can sometimes be helpful, for example, reframing a regression problem to a ranking problem that may be actually more suited to your use case and therefore the modeling techniques you use.

This is a brief overview of statistical considerations for your modeling, intended as just an introduction to put them on your radar in case you haven't considered them before. If you'd like to read more about them, check out the R community via various blogs and online books—a good place to start us the RWeekly.org (*https://rweekly.org*) newsletter. The data science–dedicated blog *Towards Data Science* (*https://oreil.ly/MNvMK*) also includes various articles on statistical subjects. Let's now move from what numbers and types of numbers you're looking to improve to helping you define how much you need to work on a problem.

Accuracy Versus Incremental Benefit

Another consideration when modeling is how much work you'll need to do to ach-ieve an acceptably accurate outcome (or precision/recall, as in Example 5-1). When performing machine learning, 100% perfect accuracy may not be the best use of your time even if it was achievable because that target represents a cost decision: it may be trivial to achieve 80% accuracy but take massive resources to increment from 95% to 99% accuracy. This means you should also consider the *incremental benefit* you'll get for increasing your accuracy, since at some point it won't be worthwhile for the bene-fit you gain.

For instance, say you can predict a user will spend an extra $1,000 next week if you send them an offer with 80% accuracy. This puts an average value on that prediction of $800. If you can increase the accuracy of that prediction to 90%, it will be worth on average $900, but if the cost of getting that extra 10% accuracy costs more than $100 per customer to achieve, you would lose money on the project.

To arrive at what is your acceptable threshold value, we need to go back to the estima-tions of project value and then project scenarios of various thresholds (80% accuracy, 90% accuracy, etc.). For example, say you predict that if you got 100% accuracy you would save $1 million a month, but you have cost considerations in resources estima-ted as in Example 5-2:

Example 5-2. In this example, there isn't a business case to go beyond 90% accuracy

```
100% Accuracy = $100,000 extra revenue
1 workday = $1000

Resources required:
80% accuracy = 1 day (GA4 native integration)
90% accuracy = 5 days (BigQuery ML)
95% accuracy = 15 days (Custom TensorFlow model)
99% accuracy = 45 days (specialized ML pipeline and model)

Costs versus incremental benefit:
80% accuracy: $80k - $1k = $79k
90% accuracy: $90k - $5k = $85k
95% accuracy: $95k - $15k = $80k
99% accuracy: $99k - $45k = $54k
```

The numbers are ballparks of what I would expect for a project, and obviously this is a bit of a contrived example, but hopefully you get the idea behind it and how it may affect your own projects.

Once you've worked out your potential goals for the data projects, you can move on to choosing what techniques to use.

Choosing Your Method of Approach

There are several types of models that you can use with your data, but some common ones I've used fall into the following categories:

Clustering and segmentation

Clustering and segmentation refer to the process of grouping together your metrics into a discrete number of groups that have something in common. Demographics are a common example: "women over 50" or "men under 30." Grouping user IDs into similar groups is a common method to start optimizing and personalizing your content. Grouping users of similar preferences can help make their experience less cookie-cutter and tailor their experiences so they will find your website more useful. It can also help predict what the customer will want if they exhibit similar behavior to similar customers in the past. For example, if customers tend to browse several widgets but then always tend to select blue widgets, you could help users by short-cutting to the blue widgets, hopefully increasing conversion rates.

Forecasting and prediction

This covers questions such as, "Where will this metric go in the future?" Forecasting can help with planning your resources. A common application is seasonal trends, such as spikes on Black Friday or over holiday periods. But other trends will be influencing your metrics as well, such as a weekly or monthly cycle. Knowing about these cycles can help you more accurately assess the impact of your campaigns. Perhaps you have a high-performing campaign, but it was only due to being at the right time of a cycle when any campaign should perform well, and the campaign did not have any inherent quality. If you were unaware that seasonal effects were giving you that false impression, you may try to copy the campaign in other less popular periods and have disappointing results.

Regression and factor analysis

Estimating a relationship between two variables is often the way to try and answer questions like, "Does my TV spend change my online revenue?" This typically looks through all the data points you're gathering and analyzes what had the most impact to your KPIs. This can be analysis that will inform later follow-on projects: for instance, you may find that rainy weather has the biggest impact on your online sales of umbrellas, so you can use weather forecasts to make sure you have enough stock to cover all orders.

Once you've selected the metric you're improving, the accuracy measure you will use to see if you're successful, and the model you're going to use, there is still one final thing to consider: how will you keep your model up-to-date when it's running in production? We'll cover this next.

Keeping Your Modeling Pipelines Up-To-Date

If you want to build your own data flows for anything beyond a one-off report, you're going to need a process to make sure you're using the most up-to-date data. You'll need to have schedules or events in place that will make the model useful on a day-to-day basis. The most common solution for this issue is to use CI/CD within your data pipelines.

This is opposed to a more traditional approach where you create an application and then deploy only at the end of its development cycle, and then changes are introduced only after another development process. CI/CD aims to eliminate delays before releasing to production, making more frequent smaller deployments rather than waiting for one big launch. This is important for data modeling in particular because things can change quickly—for instance, a new web page could be put up, and it wouldn't be introduced to your model without being added to the code or data model. A CI/CD approach means that updates will be live as soon as possible.

To enable this approach, rigorous automated testing will give you the peace of mind that your model is reliable and will work in the future. This testing must run on every change, and only upon successful completion will a push to production be enabled. This is important when dealing with data products since trust in the data is a necessity if you want to keep your users' trust. GCP embraces this methodology through various products, such as Cloud Build that we spoke of in "Cloud Build" on page 137.

The other side of the coin when deploying models is keeping an eye on their performance once live. Testing and monitoring is necessary here since data modeling can start returning poor performance even if the code is the same but the data it is ingesting contains unexpected values. To mitigate this, alerts and dashboards to inspect those results and action some change can help if you see performance fall below predetermined thresholds. It's a good idea to define the tolerance of your model performance before you deploy, with numbers you obtained when deciding what accuracy will be good enough for your model, as discussed in "Accuracy Versus Incremental Benefit" on page 172.

Advanced usage of this threshold can be using your CI/CD system to trigger retraining or modeling of your application should it fall beneath performance tolerance. In many cases it may just be that the training of the model needs to be rerun but on more up-to-date data. In that case, the model should be more self-sufficient, and you would need to get involved only if the actual model code needs updating, if a retraining does not improve performance.

Linking Datasets

Often you will need to link data across silos, as you will have identified that the answers to your questions lie in the combination of two data sources and not neatly in one. Joining datasets is usually a high value task, but sometimes it requires luck and judgment to go smoothly.

The first thought may be to link your datasets on an individual user basis, but this may not be possible or simply too complicated for a first project. Considering more coarse join keys such as campaign IDs or even joining on just the date when events occurred may be much easier to do and still surface important insights. You can still extract a lot of value, sometimes just by being able to see trends from different data sources on the same plot.

Here are some common issues when joining data:

Does a key exist?
> It may simply not be possible to perform joins on your data because the necessary granular data is not being captured or is not available. In Universal Analytics, for example, by default the `clientId` of users is not available unless you first capture it and add it to a custom dimension, meaning you had to either configure it, then wait for the data to be there or upgrade to GA360 with its BigQuery exports that included a `userId`. With GA4, this restriction eases because the BigQuery export is available. Another consideration is that you may not have a login facility on your website, so you won't have a reliable way to generate a `userId`—in that case, you may have to look at your entire website strategy if a `userId` is deemed important for your business.

Is the key reliable?
> If you have the `clientId` from GA4, it may not be a reliable method for linking users. The `clientId` is stored in a cookie, which means it can be deleted or blocked by browser restrictions, or it may not represent one person since that person will likely use several browsers (mobile and desktop). Another reason could be that a single user's browser is actually shared by lots of users. A more reliable method is to use a `userId` generated when someone explicitly logs in; but then this also means you need to configure your GA4 `userId` property and wait for the data to be there before starting your project.

Do the linking keys occur in a linked user interaction?
> If you have a GA4 `userId`, you need to be able to link that ID to the dataset you want to merge with. In practice, this means it's not enough to create a `userId` that works with only GA4. You will also need to generate that `userId` in a manner that it can be linked with your internal data. This often means that the backend server for your website or mobile application needs to generate the `userId` and

push it up to GA4. Another alternative is to enable sending the GA4 `clientId` or `userId` to the server. This can be achieved via hidden fields in HTML forms, filled in with the linking data you require when a user submits. Some CMS systems offer this as a standard, meaning you can link session information such as campaign referral data, giving you some level of actionable data without needing to do a big `userId` join operation.

How much user data will you be joining?

The nature of web analytics data is that there is a lot of messy data. Since data is flying around over unstable HTTP connections that can fail or mangle data points, your web data is never 100% reliable. There is also usually a lot of it since every user interaction is recorded. For global websites, this can add up to GBs of data per day. If the use case you're looking at is linking on a `userId` level but planning to aggregate it up again (say to campaign level for attribution projects), then you're looking at big, expensive joins performed daily. BigQuery can handle this, but it may be a big hammer for a small nut if your use case is only to see the campaigns that contributed to sales in your CRM system. In that case, capturing a `campaignId` and joining on that field will mean you have to perform smaller joins and get results more easily.

How to handle duplicates, one-to-many, and many-to-one links?

Even when using a backend system to help make more reliable joins, people are people and forget logins, share logins, and otherwise. This likely means you will need to deal with duplicates, many users being associated with one key, and one real user with a lot of IDs. You need a strategy to deal with this if it's a significant part of your dataset, usually governed by business rules appropriate for your use case.

Once you have added context to your datasets via linking them to other sources, that can be an end goal to itself, but often you will have to tease out insights from the data itself, such as averages, maximums, or counts. Beyond these simple statistics, you start looking at regressions, clusterings, and associations, which is where more advanced statistical techniques and machine learning come in. To lower the bar to getting to these insights within BigQuery, features were added that allow you to compute them within the dataset itself using SQL. This is BigQuery ML, which we'll discuss in the next section.

BigQuery ML

BigQuery ML (*https://oreil.ly/78tdS*) lets you execute machine learning models in Big-Query using only SQL. This means you can avoid extracting the data from BigQuery, which before BigQuery ML existed, you had to do by downloading your data for your models in another environment such as Jupyter notebooks or R data frames. This holds several advantages: it keep pipelines simple, it gives you the ability to run your models over all your data, and it enables data analysts without substantive machine learning training to apply simple models.

Since GA4 data comes with the BigQuery export, you can directly apply machine learning to your GA4 data with no other system setup. The results from the models will appear as another transformed BigQuery table that you can then export to the data activation channel we talk about in Chapter 6.

In the next section, we'll go over some of the specific models BigQueryML can execute that may be useful for your datasets.

Comparison of BigQuery ML Models

There are several BigQuery ML models available, and more are added all the time, but here are some pertinent examples for GA4 data:

Linear regression
Linear regression is an approach to modeling a relationship between two variables. In its basic form, you can use it to put a trend line through a group of data points. Linear regression models allow you to forecast your GA4 time-series data with the simplest model available. Don't let "simple" make you think it will be less accurate than more sophisticated techniques: it is a rule of thumb that the better the data collected, the simpler the model you need to achieve good results. Better data and a simple model will outperform complicated models and poor data. You can use linear regression models, for example, to forecast the number of items that will sell on a given day.

Logistic regression
Logistic regression is related to linear regression but allows for only binary outcomes, e.g., whether a user transacted or not. You would use this to predict whether a user will convert or not, for example.

K-means clustering
K-means clustering is a machine learning model that tries to find groups within your data points that share similarities. Clustering techniques are used to group similar data together. In a GA4 context, you could use them to identify segments of users with similar purchasing behavior. K-means is an *unsupervised* machine learning technique, so the segments grow organically out of the data. *Supervised*

techniques have you predefine what groups you want to assign your data points to. This means you can use k-means to identify how many different types of customer behavior you have, such as big spenders, one-time-only visitors, and frequent but low spending visits.

AutoML tables

BigQuery ML also integrates with some automated machine learning products such as AutoML. This lets you rely on prebuilt models that will scan your data according to the goal you wish to achieve and select the best approach for you, without requiring you to review and compare contrasting manual models. Quite often these models will quickly give you a model that will outperform your own manual creation.

TensorFlow model importing

If you have established machine learning scientists who can outperform AutoML or the prebuilt models available in BigQueryML, you can still take advantage of the in-database approach by supplying your own custom-built TensorFlow model. TensorFlow is a popular, world-leading machine learning library.

Figure 5-5 shows a decision tree that aims to help you select which BigQuery ML model would work best with your data.

The models discussed are available in many other machine learning platforms as well, but since your GA4 data will nearly always start in BigQuery, using BigQuery ML will be the most direct route to implementing machine learning models to your data. It may be, however, that you have existing data science workflows that you will prefer keeping and instead rely on exports from BigQuery.

You can use the other machine learning platforms if they have enterprise features, but the next section will look at how to deploy the BigQuery ML models in production, highlighting the key needs and solutions available.

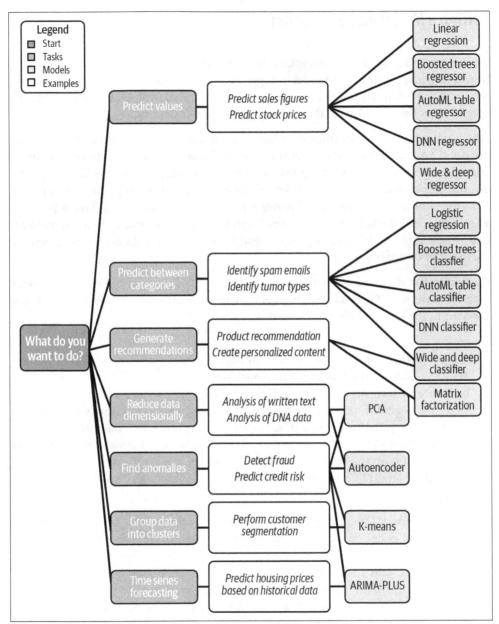

Figure 5-5. Cheat sheet adapted from the BigQuery documentation (https://oreil.ly/ e1B8G) showing what use cases and BigQuery ML model may be most appropriate

Putting a Model into Production

Once a model has been crafted from your GA4 and other data, you'll probably have a static model that works within your criteria for the data you're seeing as of today. However, when putting that model into production so it can start helping your business, you'll need to consider how that data environment will evolve over time and how your model will adapt to those changes to keep performing.

This is where the decision from your business use case will really make a difference, such as if you need real-time streams or batched predictions, since it's on that decision where the technology you consider using on an ongoing basis will diverge. As mentioned previously in "Data Modeling" on page 34, you should also have an idea what thresholds indicate a "good" performance for your model and have a plan for when the model starts to underperform. From your use case, you should also have an idea where your model data is going to end up within the data activation channels discussed in Chapter 6.

These considerations mean that you will still have work to do even once your model is performing well. GCP has many tools to help with this step. Here's a list of a few I've used in the past, to help with your projects:

Firestore

> We covered Firestore in "Firestore" on page 120, highlighting its real-time read abilities. Often, you'll want your model results to interact with users as they browse your website, for example, if they are a loyal customer and you have modeled their segments or buying preferences. Porting your BigQuery results to Firestore means you can get real-time responses for your users but still not need a fully real-time data flow since you can batch updates in between BigQuery and Firestore (for example, daily or hourly updates). The Firestore data can be reached via any HTTP application, such as Cloud Function or within GTM.

Cloud Build

> We spoke about Cloud Build in "Cloud Build" on page 137. Cloud Build deploys your model code and reacts to other events such as GitHub commits and file updates via Pub/Sub. This allows you to create pipelines that check your model performance and then trigger a retraining of your model should it fall below thresholds. In many cases, training your model on the new data will push the performance back up, but if it doesn't, you can fire off emails to have someone take a look at updating the model code.

Cloud Composer

> Cloud Composer's batched scheduling abilities make it suitable for large pipelines that typically include a modeling step. See "Cloud Composer" on page 131. Cloud Composer also frequently handles data ingestion and sending the results of your data model to your activation points as well.

GCP focuses on the ease of implementing machine learning as one of its goals, so this is an area that GCP should excel in once you have the need. To make deployments even easier, in many cases you don't actually need to make any model at all but rather can use prebuilt models available from Google via its APIs, which have been trained on richer datasets. These are discussed in the next section.

Machine Learning APIs

Many problems have been solved previously by machine learning experts, so it may not be the best use of your resources to reinvent the wheel, especially if your wheel-making equipment is not as sophisticated as the wheel factory. You may end up spending a lot of money for an inferior product. For that reason, it's very much worth checking out the prebuilt models that are available to see if they can fit your use cases.

This is a rapidly evolving area, so I'll parse out the many different AI products (*https://oreil.ly/x769N*) that will be most relevant for GA4 and digital marketing workflows. In particular, they're useful for activating difficult-to-process data such as PDFs, videos, paper records, or photographs and turning them into much easier formats such as a BigQuery table:

Natural Language AI
> This is useful for turning free-form text from social media, emails, and product reviews into structured text that picks out the sentiment, subject, category, and people and places involved. This allows you to create trends for these fields without needing to read each individual word. In particular, this is useful when combined with APIs that extract text from other sources such as video, audio, images, and translations.

Translation API
> This is a more sophisticated version of the Google Translate you may know from the web and can be used to translate various languages via an API call.

Vision optical character recognition (OCR)
> There is often a large amount of data within companies that is not yet digitized, such as paper records. Scanning these to digital photographs is the first step, but then OCR can be used to extract text from those images and turn it into structured data.

Video Intelligence
> Video intelligence is a difficult format to work with. Using the Video Intelligence API, you can extract categories and speech from the video files, which can be used later to turn its unstructured nature into structured data.

Speech-to-Text

Audio can be used within your data pipelines by turning raw audio files of speech into text formats. The Speech-to-Text real-time capabilities also offer a route to activating your data by letting users make their request via speech, similar to using Google Assistant or Siri.

Timeseries Insights API

A recently released API looks to solve the common use case of finding anomalies within time-series data, such as a GA4 session or page view count. Sending your time-series data into this API will return any unusual events outside of historic trends, which can be useful for spotting tracking mistakes or unusual user activity.

In the next section, we'll cover how to use these day-to-day.

Putting an ML API into Production

To use machine learning, APIs should be the quickest way to add it to your data flow applications with the least amount of resources or skills needed. You essentially need to make sure the import data is of the right format as mandated by the API documentation and then have a place to store the results—in most cases, BigQuery will suffice.

An example I've used in the past has been to turn unstructured text into structured data, e.g., turning a free text field into one you can put into a database with identification of the important entities or words within that sentence, sentiment, and categorization of the words within. An example of this is turning email support text into a structure format that can flag complaints versus praise or bulk rating the sentiment score of comments on products listings to catch upticks in technical problems.

This area of text analysis is known as natural language processing, and Google provides the Natural Language API to return results based on uploaded text. I have worked with it via the R package googleLanguageR (*https://oreil.ly/6mR8U*), but it's available in several other SDKs, including Python, Go, Node.js, and Java. A suggested workflow is to make it an event-based system that reacts to new files added to a GCS bucket that contains the text you want to analyze. The GCS files would be added via the host system receiving the comments, emails, or whatever you wish to process. Figure 5-6 shows this pipeline with a Cloud Function writing the results to a BigQuery table for you to use further down your pipeline. Note that we are turning unstructured data into structured data to help standardize what you're working with.

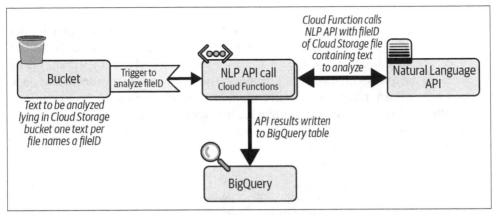

Figure 5-6. An event-based pipeline for processing text files on GCS as they arrive and putting the Natural Language API results into a BigQuery table

The Machine Learning APIs are part of a wider AI/Machine Learning platform on GCP that is specialized to popular use cases. Once you have needs beyond those, you start to get into creating your own custom models that the Vertex AI platform supports, which we'll explore in the next section.

Google Cloud AI: Vertex AI

GCP has created a dedicated machine learning infrastructure for deploying models called Vertex AI. Vertex AI allows you to deploy machine learning models that respond to HTTP requests, such as when a GA4 userId is compared against a database to see if it has features. To use a custom model in Vertex AI, you typically need to put your model coded in Python, R, or another language into a Docker container and then send that to the Vertex AI servers, or you can use one of its prepackaged models for common tasks such as regression, classification, and forecasting.

Once you are using Vertex AI, you're into advanced data modeling, which is outside the scope of this book, but I will go over some of its capabilities to give you an idea of what is possible, keeping in mind that in most cases your entry point will be the Big-Query GA4 exports:

> Be aware that Vertex AI is currently only available in certain GCP regions, so check that it's available in the same region as your data. See the Vertex AI documentation (*https://oreil.ly/vZ4A4*) for current locations per feature.

AutoML Tabular

> AutoML Tabular covers instances where the source data is a tidy rectangular dataset, such as a processed GA4 dataset. It is well suited for use on GA4 data once you've transformed the data from its raw unedited form, perhaps using some of the queries highlighted in Example 3-6. Once you have your tabular data, you can perform regression, classification, or forecasting models. An example of creating a Vertex AI dataset from BigQuery is shown in Figure 5-7.

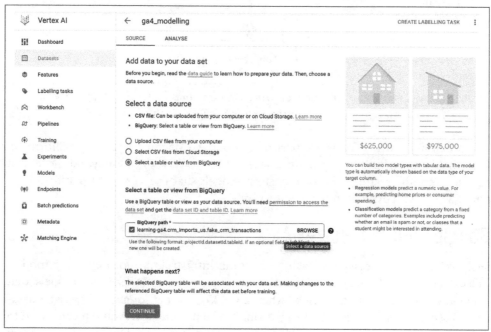

Figure 5-7. Creating a dataset from BigQuery in Vertex AI

AutoML Image/Video

> AutoML Image and Video contains machine learning models that classify your supplied images/video with labels as you wish so that when new images or video are input into the model, it will try to predict the label most similar to your training set. Since GA4 isn't collecting images or video data, you won't be able to use its capabilities directly, but you may have thousands of images and videos on your website that you wish to classify but don't have the resources to do it manually. An approach to solve is uploading those images to this service.

Workbench

> Jupyter notebooks are a common platform for developing machine learning models. Vertex AI Workbench lets you run your code within those Jupyter notebooks within a Google-managed system that has easy authentication with all GCP services, meaning you can develop your models from any computer with a

browser and do not need to run heavy graphic card processors installed on your local machine. Relevant mostly for advanced data science practitioners.

Pipelines

Pipelines are a scheduling and monitoring tool much like Cloud Composer (see "Cloud Composer" on page 131) but specialized for machine learning models. Pipelines runs your code within a Kubernetes cluster and can scale upward as you need them in a serverless manner. Relevant mostly for advanced data science practitioners.

Endpoints

Endpoints take your machine learning models across your custom or off-the-shelf applications and make them available via an API endpoint, so you're creating your own API, as we'll talk about in "Creating Marketing APIs" on page 225. This could then be your route to scaling up your models so they're available to other coding languages and services over HTTP.

Labeling tasks

You may have a lot of data that you wish to label for use within your data model but haven't got the resources to manually do it yourself. This is a critical first step if you're looking to scale up to new data, such as if you're using image or video classification in the AutoML services. Google offers a paid service for labeling tasks where you can send your data to its own workforce for them to label the data for you, following the instructions you set out in the web interface.

Putting a Vertex API into Production

The means to put Vertex AI pipelines are embedded within the platform itself, since it's a data activation product specifically for machine learning. If you're using Vertex AI, it is expected you will have some data scientists or data engineering resources to own the process and take charge of it, although a digital marketer dabbling with GA4 datasets may perhaps find enough value in using the AutoML models. The general shape of involving Vertex AI in your workflows is to have your GA4 and other datasets available in the Vertex Datasets suite, to be working on them using the prepackaged or custom models within its Workbench notepads, to orchestrate scheduled flows using Pipelines, and to make the predictions available via Endpoints. To explore this further, I suggest the reference *Data Science on the Google Cloud Platform*.

Integration with R

This section is special to me since it is with R that I have been most able to activate data within my career, and so I have the most experience with its uses. It's certainly not necessary for modeling with GA4 data; other languages such as Julia or Python will get you there, but R is a programming language dedicated to data modeling, so it has tools that make data science use cases quick to realize. The biggest issue back when I started using R was getting Google Analytics data imported in the first place, which is why I created the `googleAnalyticsR` package. Once you have data within R's ubiquitous `data.frame()` object, then it can work with the thousands of other data packages in R's ecosystem, including statistics, machine learning, plotting, and presentation, which I will cover next.

Overview of Capabilities

R is responsible for how I think about data science projects because the medium and syntax of its languages encourages certain data practices. For instance, vectors are a default, not scalars (a length 1 vector), as is the case in nondata analysis languages, meaning basic programming standards such as looping over a vector change nature when done in R as compared to other languages. Vectors naturally represent columns within a `data.frame`, so embedded within R is the concept of thinking of your data in relation to its other entries, rather than just the entry itself.

Certainly, other languages can copy the approach, as Python has popularized with its `pandas` library, but as the R community continues to innovate, new data analysis best practices do tend to start there. Python has certainly eclipsed R in uptake given its applications with deep learning such as TensorFlow and PyTorch, but I argue that unless you're involved in deep research of those tools, the R ecosystem will perform just as well. Another big plus for R is its fantastic community, which is as much a feature of the language as its technical abilities.

The steps for a data analysis project follows a certain sequence, and R has packages that can handle all of them. Selected highlights with a digital marketing and GA4 bias are shown here:

Data importing
> Various functions for importing data such as `read.csv()` exist within R, as do a host of packages to import from other sources, such as my own `google AnalyticsR` for GA4 data. There is also a dedicated Web Technologies Task View (*https://oreil.ly/Vqzp5*) that lists packages for importing from popular websites such as Facebook, Search Console, and Twitter. There are also many database connection libraries that have a common framework with the Database Interface (DBI) package (*https://oreil.ly/hkbFI*) to allow connection to BigQuery, MySQL, or other databases as necessary.

Data transformation
> The tidyverse is a suite of packages that have inspired me very much and is the implementation of the tidy data practices I talked about in "Tidy Data" on page 102. Packages such as `dplyr` give you an easy-to-read, piped data flow that is data source agnostic in that it can translate its R syntax into SQL. `tidyr` carries

various tools for easy shaping of data into tidy versions of themselves, and `purrr` contains some iterative vectorized looping functions that make working with nested columns or other processing jobs easier. Special thanks to Hadley Wickham of RStudio for their creation.

Data visualization

Base R contains many handy functions for quick plotting of data, which is essential when carrying out analysis to make sense of what you're doing quickly via its function `plot()`. You can also create passable visualizations for public consumption using `plot()`, but if you would like power and flexibility, `ggplot2` is widely recognized as the professional standard. `ggplot2` implements a "grammar of graphics" that changed the way I think about visualization and the shape of my data.

Data presentation

A key part of a data pipeline is producing results that can be read and digested by others. Interactivity is also popular to encourage consumers to explore the data on their own to a limited degree. Two very useful applications in R are Shiny (*https://oreil.ly/8TscS*) and R Markdown (*https://oreil.ly/o7PSM*). Shiny is a syntax to create online data applications that run R code when users change app settings via drop-downs or input fields. It is a highly polished product that can create applications quickly and can be customized to look like any web page. R Markdown is a subset of Markdown that can run code chunks and display that code or just its results as it is being built with your choice of format. It essentially changes R code into professional-looking PDFs, HTML, or Word documents, and in my mind it is superior to the more popular Python Jupyter notebook format for presenting data science work. Because R Markdown can render to HTML with JavaScript, it can also maintain some level of interactivity even when not running R code, and can be used to create websites. For instance, I used `blogdown` to create my blog (*https://oreil.ly/zRok0*), which is a package that is built on top of R Markdown. Now, you don't even need to run R to use it, since Quarto (*https://quarto.org*) has been released—that has taken its lessons from R and made it available for all languages via its standalone dedicated tool.

Data infrastructure

Using a data science language means that qualities such as reproducibility and scale are highly regarded, so there is a whole set of meta packages within the R community that also deal with the meta work of producing high-quality data workflows. `targets` is an example package that allows you to run or skip data pipeline steps as necessary but in a reproducible way. This means you can prevent large unnecessary data steps if only minor code changes have taken place. It recognizes that data analysis work often involves minor changes and corrections when iterating on your analysis. I also mention my own `googleCloudRunner` and `googleComputeEngineR` packages, which provide tools for working with Docker images of your R code to run it on a VM or serverless environment within the Google Cloud.

Data modeling

In keeping with the theme of this chapter, R has an extensive number of statistical and machine learning packages to extract information out of your raw data. This is its major strength, since even in its default installation it contains many forecasting and clustering models with its own syntax embedded within the language. The statistical capabilities of R are usually why R is considered in the first place.

Despite all of this, there are reasons why R is not the number-one choice for all data analysis. It is not made for hardcore developers but for statisticians and end users, so some of its quirks can be tough to swallow for developers coming from other languages. It also both benefits and suffers from being regarded as a domain specific language (DSL) for statistics since more general-purpose languages such as Python mean less context switching for the developer. It's also an open source language, so versioning of packages and dependencies can be hard to navigate, although I argue that R's CRAN package system is much more robust than other language module systems because every package is at least reviewed by one human being before publishing. However, many of the reasons mentioned have led some to conclude that "R cannot be used in production," which I know to be false as I have many R scripts in production. I think the easiest way to do this is by using Docker, which I talk about next.

Docker

Docker is described in "Containers (Including Docker)" on page 44, and I regard it as a useful tool for data analysis in general. This section discusses it for R specifically because I regard it as the best way to use R in production.

Docker addresses perceived weaknesses in using an open source language in production since it enables you to pin down the exact environment R is running in. You don't need to worry about updates or package dependencies because the code you're running with a Docker environment is sandboxed, secure, and isolated from the changing bleeding edge of the software development. Using Docker also means you can share the results of R code with non-R users, and they can benefit from its output without needing to know R itself.

Docker is also used extensively within GCP as the mechanism to provide custom code solutions. Within Google's Vertex AI, you have the option of both using its pre-packaged solutions and supplying your own custom code via a Docker container, which could run R, Python, Visual Basic, or any language you can fit in a Docker container. What works well with R should work well with any data analysis language, so an investment in learning Docker should never be wasted.

Using R in Docker is greatly helped by the Rocker project (*https://www.rocker-project.org*), an open source initiative that provides multiple useful images with R pre-installed. Images include a specific version of R: images with RStudio is the most popular R IDE preinstalled; images with the tidyverse, as mentioned in "Tidy Data" on page 102; and also GPU-enabled images with machine learning libraries such as TensorFlow and PyTorch installed.

Using Docker is an essential component when using R in production, an example of which appears in the next section.

Example 5-3 shows a Docker file for running R. It assumes you have a self-contained R script running in the directory with a script for your report at scripts/run-report.R. All the libraries and systems dependencies are installed, and then the script and any other files within the same folder (such as configuration files) can be loaded in for a self-contained container image. The ENTRYPOINT command executes by default when running the container—in this instance, the R script.

Example 5-3. Using Docker to run an R script

```
FROM rocker/tidyverse:4.1.0
RUN apt-get -y update \
 && apt-get install -y git-core \
        libssl1.1 \
        libssh-dev \
        openssh-client

## Install packages from CRAN
RUN install2.r --error \
 -r 'http://cran.rstudio.com' \
 remotes \
 gargle \
 googleAuthR \
 googleAnalyticsR \
 ## install Github packages
 && installGithub.r cloudyr/bigQueryR \
 ## clean up
 && rm -rf /tmp/downloaded_packages/ /tmp/*.rds

COPY [".", "/usr/local/src/myscripts"]

WORKDIR /usr/local/src/myscripts

ENTRYPOINT ["Rscript", "scripts/run-report.R"]
```

The Docker image is typically built and pushed to Google Artifact Registry, so it's available for further-downstream applications. This may be a Cloud Build step or Cloud Composer DAG. If you want to run your R script in response to HTTP requests, such as with Cloud Run, you will need to have a way for your R code to respond to an HTTP request. This is usually done with the R package plumber, which includes a syntax to have your R code respond and request data from HTTP.

R in Production

There are several ways to work with R in production on GCP. I'll give an example showing R within a batched Cloud Composer schedule using the Docker image shown in Example 5-3.

Take the DAG example written in Python that will call the Docker image. It will use Airflow's KubernetesPodOperator to start up the Docker image and run it. In the example shown in Example 5-4, the R code includes a script for downloading GA4 data and uploading it to BigQuery. An authentication file is kept securely in a Kubernetes secret since it's not a good idea to embed these within Docker containers because they are then available for anyone who runs it.

Example 5-4. Using the R Docker image within an Airflow DAG

```
import datetime
import os
import logging
from airflow import DAG
from airflow.providers.cncf.kubernetes.operators.kubernetes_pod import(
        KubernetesPodOperator)
from airflow.kubernetes.secret import Secret
from airflow.providers.google.cloud.operators.bigquery import BigQueryCheckOperator
from airflow.utils.dates import days_ago

start = days_ago(2)

default_args = {
 'start_date': start,
 'email': 'me@email.com',
 'email_on_failure': True,
 'email_on_retry': False,
 # If a task fails, retry it after waiting at least 50 minutes
 'retries': 3,
 'retry_delay': datetime.timedelta(minutes=50),
 'project_id': 'ga4-upload'
}

schedule_interval = '17 04 * * *'

dag = DAG('ga4-datalake',
   default_args=default_args,
   schedule_interval=schedule_interval)

# a Kubernetes secret used to store an authentication file
secret_file = Secret(
```

```
 'volume',
 '/var/secrets/google',
 'arjo-ga-auth',
 'ga4-import.json'
)

# https://cloud.google.com/composer/docs/how-to/using/using-kubernetes-pod-operator
arjoga = KubernetesPodOperator(
 task_id='ga4import',
 name='gaimport',
 image='gcr.io/your-project/ga4-import:main',
 arguments=['{{ ds }}'],
 startup_timeout_seconds=600,
 image_pull_policy='Always',
 secrets=[secret_file],
 env_vars={'GA_AUTH':'/var/secrets/google/ga4-import.json'},
 dag=dag
)
```

Note that this Docker image could also be used in other systems such as Cloud Build, which demonstrates one of the most powerful aspects of using R scripts within Docker: you can switch it to other systems, even to other clouds, easily.

Summary

It's impossible to go very deep into all the possibilities of data modeling in just one chapter of this book, so I've tried to demonstrate starting points that begin with your GA4 data and invite you to take it further. This part of data pipelines is a rich tapestry of possibilities, and you can build a whole career on refining which and what models, tools, and frameworks you wish to use. In this chapter, I have tried to select technologies that can be actioned the quickest to get your results with your GA4 data, which, because it sits within BigQuery, has an easy route to results via BigQuery ML. Should you then need more on top of BigQuery ML, the machine learning APIs open up a rich set of new possibilities. In addition, the brief overview of Google's Vertex AI should show you that the ultimate limit of your modeling can be super-sophisticated models that any data scientist would be happy to work with. Next, we'll dig deeper into data activation.

## CHAPTER 6

Data Activation

Data activation is the business end of the project, where we expect to generate return on investment, impact, and value. When we talk about data activation in this chapter, we embrace many different applications, but the most common definition is when you are able to use your data to inform business decisions or change user behavior. Without the ability to change anything, your data project will have no influence and may as well not exist. Your data project can create influence in many ways: it could be a one-off insight that the CEO keeps in mind when allocating budgets, a day-to-day metric tracker data analysts use to decide where to work next, or an autoupdating website feature that adjusts prices or content automatically. All can be regarded as data activation, but some will be harder to measure or have less impact than others.

Because of this, data activation can be regarded as the most important consideration if you're looking to make data projects beyond only educational reasons or a proof of concept. You should at least outline how you're going to activate your data when you're scoping out the project, so we go into how to keep this a priority in the next section.

GA4's data activation integrations are most likely the key reason to use GA4 over other analytics solutions, especially if you're using other Google digital marketing services in your business online, such as Google Ads. It is a key differentiator for the product and one of the prime reasons Google Analytics is offered for free: Google knows that the more your business can measure the performance of your digital marketing campaigns, the more likely you will increase budget on those Google Ad services. Most of these data activation features are enabled via Audiences, where you can segment users into buckets and export those attributes to services such as Google Optimize, Search Ads, or Google Ads.

Importance of Data Activation

Data activation can sometimes be neglected and overshadowed by data modeling, but I now regard it as the most important aspect—a poor model with good activation is better than a good model with poor activation. A clue that you may not be properly considering data activation is if you only start to think about it after the data modeling stage or you have assumed a dashboard will be the end result without questioning that assumption. This section will introduce some concepts that will help you decide what is best for your own projects.

As stressed in Chapter 2, when you're planning out your data projects, you should have a clear idea of how much benefit to your business that project will theoretically achieve, and this usually will come down to estimations on how much value the additional data activation phase will add to your business. This usually comes from a cost saving or by providing extra revenue. Some techniques to estimate these figures include the following:

Time savings via efficiency
> A common goal is to help automate a certain action your colleagues are doing that could be optimized via an automated service. An example is pulling all metrics into one place so a user need only log into one place to get the information they need, rather than needing to spend a few hours each week logging into every service, downloading data, and aggregating the data by themselves in a spreadsheet. You could then arrive at a cost-saving number by estimating monthly hours saved multiplied by the average hourly rate of those users.

Increasing ROI performance of marketing costs
> With GA4's digital marketing focus, a common need is to increase your conversion rates or click-through rates by providing a better website or advertisement experience via improved relevance for that customer. If this increase in rates can be attributed to your project, you can work out an incremental increase or revenue given typical monthly volumes of traffic.

Reducing costs of marketing
> A similar activation strategy may be trying to target the same number of customers more efficiently so you spend less to attract the same number. A common technique is to tailor the keywords you're targeting via paid search or to geotarget those users and exclude customers you think will never buy (perhaps they are customers already?). You could then attribute the reduced incremental costs per month to your data project.

Reducing churn of existing customers
> Some data projects are justified by increasing customer satisfaction so you can increase the number of repeat buyers and reduce churn. You can do this via personalization or by identifying annoying sales patterns you can exclude your

existing customers from. It is said that the cost of acquiring a new customer can be 10 times the value of keeping an existing one, so you can look to attribute this cost or the incremental increase in revenue from repeat buyers to your project.

Attracting new customers

Most businesses also need a regular intake of new customers, so discovering new customer segments outside of your existing base can carry high value, especially for growing startup companies. Creating look-alike audiences that identify possible customers similar to your existing or keyword research for users looking for similar products could be a motivation for your data project. Competitor research may also feature here. You could then attribute any uplift of new customer signups to your data project.

In many cases, how much value you can add via your data activation will be an educated guess, but having a ballpark figure is still important so you can compare your expectations to reality and, of course, get sign-off on any budget you may need for the project. It should also help you lock down what resources you will need and what data is necessary, and you'll find that this is also where you will interface the most between the business and technical areas of your company.

You may decide a dashboard is the best activation channel, but "Visualization" on page 203 considers some caveats to that assumption, to make sure dashboards perform as you expect.

GA4 Audiences and Google Marketing Platform

A major reason Google Analytics is a preferred solution for many companies is its tight integration with Google Ads and the rest of the Google Marketing Platform (GMP) (*https://oreil.ly/Q8MJj*). GA4 is in a unique position because of its integration capabilities versus other analytics platforms, since Google Ads is, for many, their most important channel for digital marketing.

The GMP includes these solutions and roles:

GA4

The subject of this book. Website and mobile application measurement and analytics.

Data Studio

A free online data visualization tool that can integrate with many Google services, including Google Analytics and BigQuery. Used often to make a presentation layer on top of your GA4 data combined with other sources.

Optimize

An A/B testing and personalization tool on your website. This tool can change what content website browsers see and then record their activity to see if goals such as conversion rates are improved via its statistical modeling.

Surveys

Creates online survey tools that pop up on your website so you can gather qualitative data from your users to supplement the quantitative data from your analytics.

Tag Manager

Covered in this book in "GTM Capturing GA4 Events" on page 53. This is a JavaScript container you place on your website so you can control all the rest of your JavaScript tags from one centralized location without needing to update the website each time. Includes useful triggers and variables commonly used by analytics tracking, such as scroll and click tracking.

Campaign Manager 360

A centralized digital media management tool used by advertisers and agencies to control when and where to serve digital ads.

Display & Video 360

Used by businesses looking to advertise on video and display networks. Helps users design advertisements, purchase them, and optimize campaign performance.

Search Ads 360

Used by businesses looking to advertise keywords in search engines, including Google Ads, Bing, and MSN.

All of these platforms could be considered data activation channels, apart from GA4 and Surveys, which are more data ingestion channels. The key selling point of the GMP is that audiences can be created in GA4 and then exported (if you have consent) to the other services.

This means that data you collect with GA4 such as user preferences can be used to influence what media they see on other channels, such as video or search. The next section goes over how to set up these audiences within GA4.

Audiences are a GA4 feature that lets you combine the metrics, user properties, and dimensions you collect into grouped buckets or segments with similar values. Their first use is to help aid analysis, such as identifying all people who purchased or viewed a particular piece of content. You can add many criteria to create very particular audiences. These audiences gain a lot of power when expanded to other services since they can then be used to tailor content or website behavior for just the subset of users.

You can take a look at some existing Audiences used in these examples via the demo account for GA4 (*https://oreil.ly/fpQiY*) for the Google Merchandise Store, found via the menu in Configure → Audiences. Several with various criteria are listed, which we can see in Figure 6-1. From here, we can get a glimpse of the various types of audiences possible.

Audience name	Description	Users ⑦	% change	Created on ↓
✨ I/O 22	Users who are predicted to generate the most rev...	1,803	-	Apr 18, 2022
✨ I/O 2022	Users who are predicted to generate the most rev...	3,101	-	Apr 11, 2022
Test Audience		1,969	-	Apr 8, 2022
Session Start and more than ...		34,768	↑238.8%	Mar 21, 2022
Session Start >>> Viewed App...		5,150	↑15.0%	Feb 1, 2022
testaudtrigger		< 10 Users	-	Jan 19, 2022
✨ Predicted 28-day top spenders	Users who are predicted to generate the most rev...	4,729	↑36.0%	Jan 12, 2022
Untitled audience		< 10 Users	-	Oct 21, 2021
(Session Start >>> Viewed Ap...		18,552	↑9.6%	Sep 30, 2021
Add to Cart		8,777	↑31.8%	Sep 15, 2021
✨ Likely 7-day purchasers	Users who are likely to make a purchase in the ne...	7,268	↑8.8%	Aug 24, 2021
✨ Likely 7-day churning users	Active users who are likely to not visit your proper...	799	↓38.3%	Aug 20, 2021
Android Viewers	Those that have viewed Android products	1,592	↑18.8%	Nov 4, 2020
Campus Collection Category ...	Those that have viewed the campus collection ca...	1,263	↑20.3%	Nov 4, 2020
Engaged Users	Users that have viewed > 5 pages	19,853	↑22.1%	Oct 5, 2020
Added to cart & no purchase	Added an item to the cart but did not purchase	8,527	↑31.3%	Sep 17, 2020
Purchasers	Users that have made a purchase	1,960	↑9.1%	Sep 17, 2020
Users in San Francisco	Users in San Francisco	1,204	↑28.8%	Jul 31, 2020
Recently active users	Users that have been active in the past 7 days	66,593	↑16.0%	Jul 31, 2020
All Users	All users	88,631	↑15.1%	Oct 19, 2019

Figure 6-1. A list of GA4 Audiences taken from the GA4 demo account for Google Merchandise Store

Included within Figure 6-1 are different types of audiences we will focus on:

Session Start and more than two page views

This is an example of an audience created by counting event parameters. Using custom audiences, we look to include users who have a `session_start` event (e.g., arrive at the website) and then have `page_view` events where Event count > 2. This is kind of like an enhanced engagement segment. Figure 6-2 shows a possible configuration for this Audience. It's simple to configure for three or more `page_views`—just change the value of the Event count condition. You can also use any event aside from `page_view`, such as a purchase or custom event.

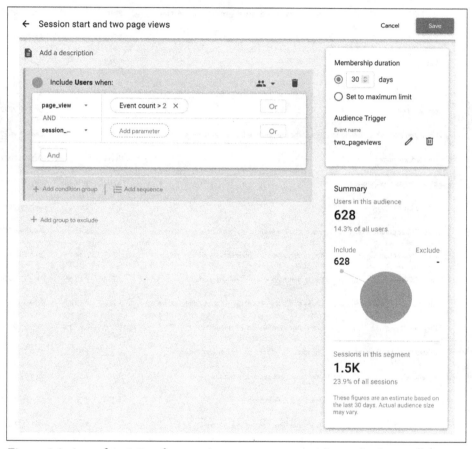

Figure 6-2. A configuration for `session_start` events with two further `page_view` events

Note that we can also trigger an additional event when a user becomes a member of that audience; this could be used in additional segments or for other measurement purposes.

Added to cart and no purchase

This is an example of a segment that includes one set of users but excludes another. The Google Merchandise Store is interested in the segment of users who looked like they were thinking of buying but eventually didn't. A good target for advertising? To achieve this, we include all users who added to a cart but exclude all users who made a purchase. Another example of this type of audience is shown in Figure 6-3, this time for users who got a mobile web notification but didn't open it. The Audience can include or exclude according to any event you are collecting. This can also include time window limits, such as if they didn't read the message within 5 minutes or 30 days.

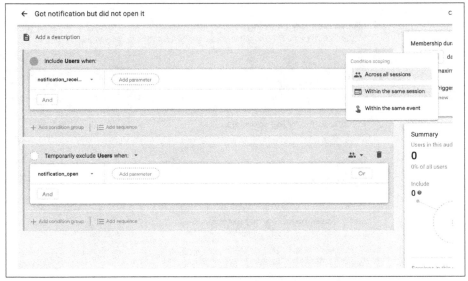

Figure 6-3. A configuration for users who received a notification but did not open it

Predicted customer

New to GA4 is the ability to not just use the data you collect but also to act upon the trends that data predicts. These are enabled via predictive metrics, which give estimates on how many customers will convert or purchase in the future based on the trends of previous customers. This is a powerful feature and can help you try to influence your customers' purchasing decisions. You can use them only if you have enough data for the predictions to be accurate, but once you do, you can find them in the Predictive menu option shown in Figure 6-4 from the Audiences menu. You can use these predictive metrics within an Audience, giving you a segment of users who may perform that action. These can be used to target or exclude those customers from digital marketing activity. Several of these Audiences are available in pre-created templates such as those shown in Figure 6-4.

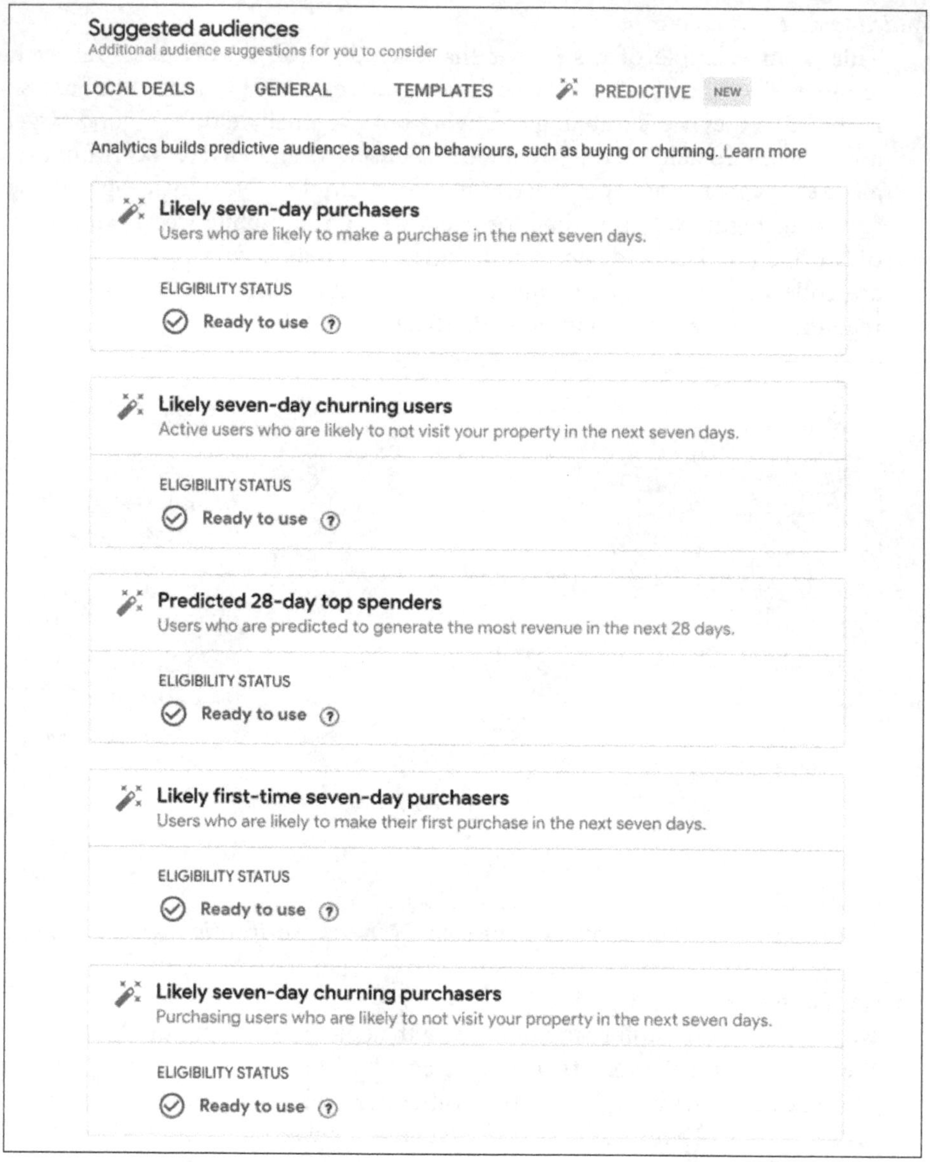

Suggested audiences
Additional audience suggestions for you to consider

LOCAL DEALS   GENERAL   TEMPLATES   PREDICTIVE NEW

Analytics builds predictive audiences based on behaviours, such as buying or churning. Learn more

Likely seven-day purchasers
Users who are likely to make a purchase in the next seven days.

ELIGIBILITY STATUS
Ready to use ⑦

Likely seven-day churning users
Active users who are likely to not visit your property in the next seven days.

ELIGIBILITY STATUS
Ready to use ⑦

Predicted 28-day top spenders
Users who are predicted to generate the most revenue in the next 28 days.

ELIGIBILITY STATUS
Ready to use ⑦

Likely first-time seven-day purchasers
Users who are likely to make their first purchase in the next seven days.

ELIGIBILITY STATUS
Ready to use ⑦

Likely seven-day churning purchasers
Purchasing users who are likely to not visit your property in the next seven days.

ELIGIBILITY STATUS
Ready to use ⑦

Figure 6-4. If you fulfill the criteria, you will see Predictive Audiences available in your GA4 configuration

Once you have crafted your Audience within GA4, activation occurs by exporting that Audience to your choice of GMP service. One of these is Google Optimize, which we will look at next.

Google Optimize

Google Optimize is a website testing tool where you can serve up different content to different users to see which performs best. It allows you to test hypotheses about how your website can perform better. For example, perhaps you suspect that the red "Add to Basket" button on your website is confusing customers who aren't used to that color for that button role. Perhaps changing the button color to green will mean more conversions, but you don't want to lower your revenue accidentally if you're wrong. Using an A/B testing tool such as Google Optimize means you can test these two variations side by side by serving some customers one variation (A) and some the other variation (B). Comparing the performance of the two should give you data on which is the best choice. Google Optimize allows you to alter your website's appearance temporarily to test these variations and makes sure a visitor always sees the same variation. The capabilities also extend to being able to serve up certain content to certain audiences or segments, including your Google Analytics audiences defined in GA4.

Previously, you could export your Google Analytics segments when using Universal Analytics, but you had to be a paying Optimize 360 customer. GA4 removes this barrier, opening up personalization, experimentation, and on-the-fly alterations of your website to everyone based on the Audiences you define in GA4.

Once you have linked your Google Optimize and GA4 accounts, you will need to install another JavaScript snippet (*https://oreil.ly/kWNlz*) to control the content Google Optimize will show on your website, and then link your GA4 account.

Once installed and linked, your GA4 Audiences will start to appear in Google Optimize. After you've created your website content, you will have the option to select who should see it, and this is where you will be able to select your GA4 Audiences, as shown in Figure 6-5.

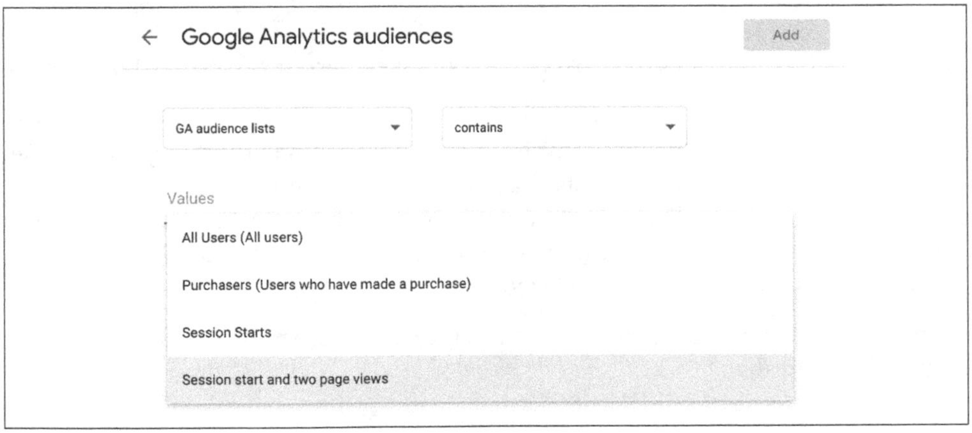

Figure 6-5. Selecting a GA4 audience within Google Optimize

I applied one of the several choices you have for your Optimize activation. This includes running A/B tests, changing content, and redirecting users to another page, but my choice was to display a banner at the top of the website to demonstrate that the segment worked, in case you may visit my blog (hint, hint). If you ever did visit the website, you should see something similar to Figure 6-6.

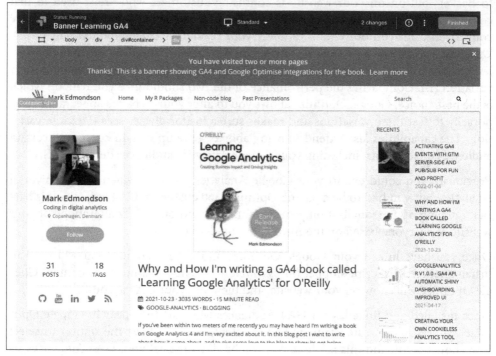

Figure 6-6. Setting up a banner for the website that will trigger when the GA4 Audience segment is fulfilled

At the time of writing, it's still in beta, so it can take a day or two to see the changes seen on your website.

This proof of concept is aimed to inspire you to leverage the Google Optimize GA4 integration for a more pertinent use case for your website. Being able to directly change your website via GA4 events in a relatively easy manner encourages experimentation with the user journeys on your website and is a good example of data activation: your data will change user behavior.

We next cover some data visualization tools since this is often the first stop along the road of data activation.

Visualization

Data visualization is the process of plotting or extracting information from your data to help inform decision making, monitoring, or trend prediction for the dashboard users. It is a vast area with many excellent textbooks covering the fundamentals, so this section will concentrate more on how GA4 enables it and some of the tools you may want to use along the way.

I was initially an enthusiastic creator of dashboards at the start of my career but then entered a phase I dub my *Dashboard Winter*, where I told everyone that dashboards were actually pointless. My sulking was mostly due to monitoring usage of the finely crafted dashboards I had spent time on. From initial daily use, the metrics soon depressingly dropped down to zero, despite the work to create them still breaking the back of many data workers.

I have retreated a little from recommending never using dashboards for data activation—I think they are a good starting place if handled in the right way. But I would also like to challenge their use and not accept them as the default, which seems prevalent in nearly all businesses, especially when we talk about activating data. I cover some of the issues with dashboards in the next section, where you can hopefully learn from some of my mistakes.

Making Dashboards Work

Dashboards come with an underlying assumption: the viewer of the dashboard will inspect the data and have an insight that will then be followed through on and used to inform a data-driven decision within the company. This assumption is not easy to achieve and should not be presumed. It requires a few things:

The right data is sent to the dashboard
> This is the technical workflow of getting data into the dashboard in the first place. It can vary from simple to complicated depending on the number of data sources it's pulling data from. This is often naively thought of as the only major task for data activation, but keep reading for other considerations.

The data is relevant at the time it is being viewed
> A common dashboard project starts with scoping with the intended user about what they will be using the dashboard for. However, in most businesses this is not static, so what you have scoped at the beginning may be irrelevant by the time your project is finished. This can often be seen by the frequency of logins to your dashboards that sharply decline over time. A potential workaround is to make the dashboard more interactive or an analysis tool so that there is some element of self-service an end user can do to keep the data relevant.

The data is presented in a clear way that the user easily understands

This is a deep and complex field that draws on design, UX, and the psychology of data interpretations. It is quite common for two people to look at the same data and draw opposite conclusions because they are informed by their own viewing context. Keeping dashboards focused and simple should be a driving force for most, but this sometimes collides with the previous point where users try to keep the dashboard relevant for all cases by putting lots of data points on the screen at once.

The data is trusted by that user

It doesn't take much for end users to stop believing the data is something they can rely on, even if everything is perfect from a technical perspective. It can take just a couple of timeouts, data processing mistakes, or incorrect data inputs to make the entire project worthless. In some cases, it may just be that the data presents answers the end users don't like. This can only really be solved with a lot of communication and by making your data pipelines as robust as possible.

The user has enough agency to act on that data

A data analyst may have the perfect dashboard, but if they can't influence their boss or other stakeholders about the conclusions drawn from it, it will never impact the business bottom line. Picking the right stakeholders to create your data products for is a key scoping requirement.

If you can assure yourself that all of these criteria can be satisfied, you are probably safe to create your dashboard, but still aim to review it periodically to keep it all relevant. For your data visualization needs, we'll consider the options within GA4 itself, then look at some of the more advanced options offered, including Google Data Studio, Looker, and other providers.

GA4 Dashboarding Options

GA4 comes with its own visualizations when you log into the tool, and to some this may be the only way of interacting with its data. I tend to spend maybe 20% of my time working with GA4 data in these reports, which I usually use only to verify data is coming in for collection. Otherwise, I'm working with its data streams via the API, BigQuery export, or GA4's various integrations.

I think the fact that people traditionally rely on Universal Analytics' web interface explains some of the reluctance to switching over to GA4, since many Google Analytics users have unconsciously learned the sometimes convoluted ways to get to the data you need. With the blank slate GA4 presents, these routes need to be relearned, and in early days of GA4, some of these reports were simply unavailable (I'm looking at you, Landing Page reports! Now implemented, thankfully). Learning a whole new GA4 reporting interface with unfamiliar routes led to early frustrations and a feeling that GA4 wasn't equal to Universal Analytics. A workaround for this unfamiliarity for new business users may be to switch entirely to Data Studio or other visualization tools and import the GA4 data into those, giving only the more technical users access to GA4 WebUI. However, that would mean missing out on many new innovations within the GA4 web interface that were impossible to do with Universal Analytics.

There are two different reporting models within the GA4 web interface that can cause confusion when comparing across them. It's helpful to regard these as two different buckets of data presentation, with differing rules on interpretation. Standard Reports are accessed from the Reports tab, whereas explorations are accessed from the Explore tab. Standard Reports provide overall aggregates for simple reports, but they lack segmentation or filtering. Exploration Reports have more analysis features such as segmentation, filters, funnels, and pathing, but may suffer from sampling. Read about these differences in "GA4 Data Differences Between Reports and Explorations" (*https://oreil.ly/Jm73b*).

GA4 Reports

GA4's "Reports" section gives you top-level overviews that aggregate the event data you are sending in a daily trend, for example, as shown in Figure 6-7. They are distinct from the Exploration Reports.

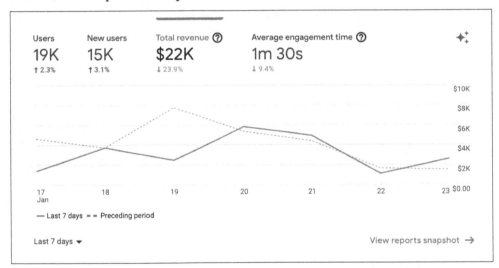

Figure 6-7. GA4 Standard Reports show you real-time updates and trends for your GA4 event data

The reports shown within the Reports section are customizable via the Library section at the bottom of the menu bar. Over time, new reports should appear that you can customize for the end users logging in. In practice, this should help you limit Reports to only those relevant for the person logging in. A valid criticism of Universal Analytics was the sheer number of reports presented to a user on first login, a bewildering experience for new users. Limiting the reports a user can see covers some of the same functionality as Views in Universal Analytics, restricting access to some data reports. By default, you should have the choice of the following reports:

Real-time reports

This data is suited to actions you are working on that day and you wish to see their impact on your website within the last 30 minutes. This could be a social media post, a marketing launch, or a tracking setup deployment. The real-time dataset is much richer than in Universal Analytics. You can add comparisons on the fly if your campaign is targeted at a particular group, or click through to user snapshots to see the behavior of a particular user. This data is also available in the Data API in real time.

Acquisition reports

This data is about how users reached your website. The key difference from Universal Analytics is that you have both user acquisition and traffic acquisition, corresponding to first-touch and last-touch channels a user has arrived from. You can also add a secondary dimension to your reports, including Landing Page (the first page that session saw), as shown in Figure 6-8.

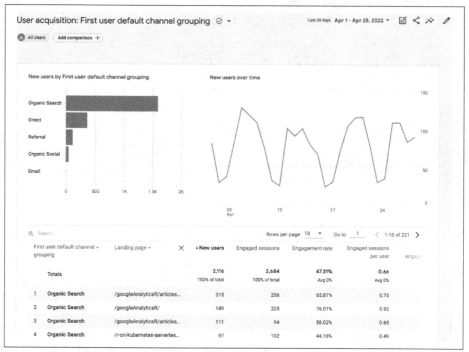

Figure 6-8. Showing how users first arrived to my blog—organic search has been kind to me!

GA4 also lets you select which conversion that particular session contributed to by changing the drop-down under your "Event Count" or "Conversions" columns, as shown in Figure 6-9.

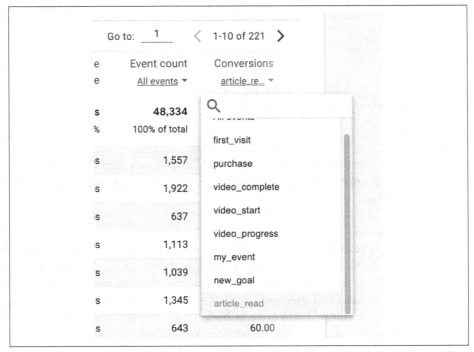

Figure 6-9. Selecting which of your events or goal conversions that channel contributed to

Engagement reports

These reports are about what events are being triggered on your website. Here you can examine the individual events you're sending in, along with their parameters. It's also where to look for page metrics similar to the All Pages report in Universal Analytics. You can also see how much a page has contributed to a goal by following the same procedure as in Figure 6-9. Events are the most granular data point within GA4. A useful report includes comparing conditions of users who viewed a certain event, such as googleanalytics_viewers, that I have implemented for those viewers on the blog, as shown in Figure 6-10.

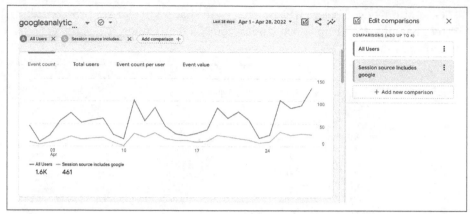

Figure 6-10. Comparison of the Audiences "All Users" and "Session Source includes Google" with a metric counting how many `googleanalytics_viewers` *events were present for each*

Monetization reports

If you are an ecommerce website, then revenue and other metrics are available. Product analytics rates such as add-to-cart, cart-to-view, and purchases all rest here, but you're more likely to want to use the Explore module for reporting things like funnels and user journey paths, leaving this report for overall totals and rates. In addition, if you are a publisher running ads, you will also find the ad-per-page revenue figures in this section. We can also see some of the new metrics such as lifetime value here.

Retention reports

Retention reports are for cohort analysis metrics, such as new and returning users and users who come back within 7 days, 14 days, etc. In the screenshot of the Google Test GA4 account shown in Figure 6-11, we can also see it flagging spikes in traffic to help indicate where you may want to implement some deep-dive analysis.

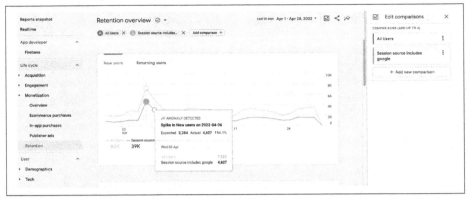

Figure 6-11. An anomaly spike found in the merchandise data within the Google Merchandise Store demo GA4 account

Demographics reports

These reports carry details about your users, such as country of origin and language settings. If you have an opt-in for more advanced demographics provided by Search Ads 360, you can also see age, interests, and gender estimations here.

Tech reports

These cover the technical details of the devices users are browsing your website or app with, such as desktop or mobile, browser, and screen resolution.

Firebase report

Firebase has various reports that are helpful for monitoring your mobile app, such as crash rate for users and your app version.

GA4 has a Library Collection feature that you can use to put together a custom list of the reports you want to see most often. For example, for my blog, I'm mostly concerned with where traffic comes from per landing page, search queries via the Search console integrations, and engagement metrics.

Using the Library functionality, I make a custom "My Blog" section in my GA4 Reports, as shown in Figure 6-12. The available reports are listed on the right, and I drag across which I would like to see when I log in to the left. There are various report types, such as an overview with summary statistics or details listing more dimensions. Once I've selected the reports I wish and have named the custom section, it will appear on the left of the GA4 main web interface within the Reports section.

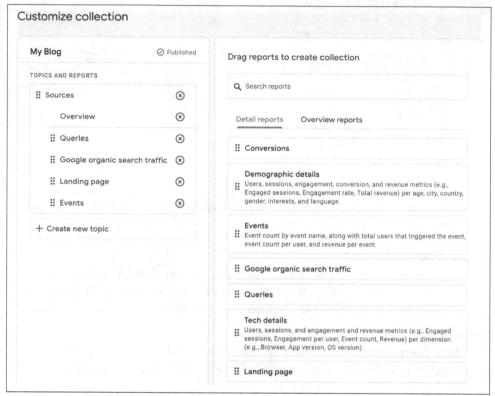

Figure 6-12. A custom collection of GA4 reports for my blog

Reports are good for getting overviews of your website performance, but sometimes you want to explore more deeply and manipulate data more. If you can't get the insights you want from GA4 Reports, consider moving to the Exploration module, which we cover in the next section.

GA4 Explorations

GA4 Explorations are accessible when you log into GA4 via the Explore menu at the top right. They are suited to more ad hoc data exploration reports and use tools such as sorting, drilldowns, filters, and segments. You can also use them to create the GA4 Audiences used in other GMP services as discussed in "GA4 Audiences and Google Marketing Platform" on page 195. The intended workflow for using them goes something like this:

1. Create Exploration: Create or select an existing report or template Exploration, such as the reports that come by default. This is the context for the use case you want to analyze. You can see the start screen in Figure 6-13.

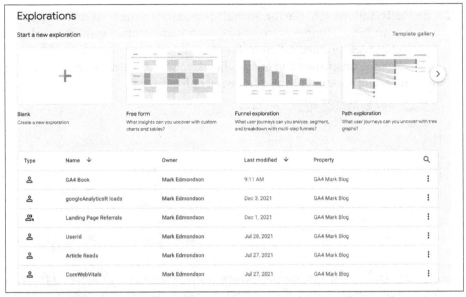

Figure 6-13. The start of your Exploration workflow involves selecting or creating one from the start screen

2. Select Variables: In the Variables section, click the + button to add or remove the relevant segments, dimensions, and metrics you think you will need. This allows you to focus on only the ones you require and avoid information overload, as can be seen in Figure 6-14. You can always modify these fields if you need to later.

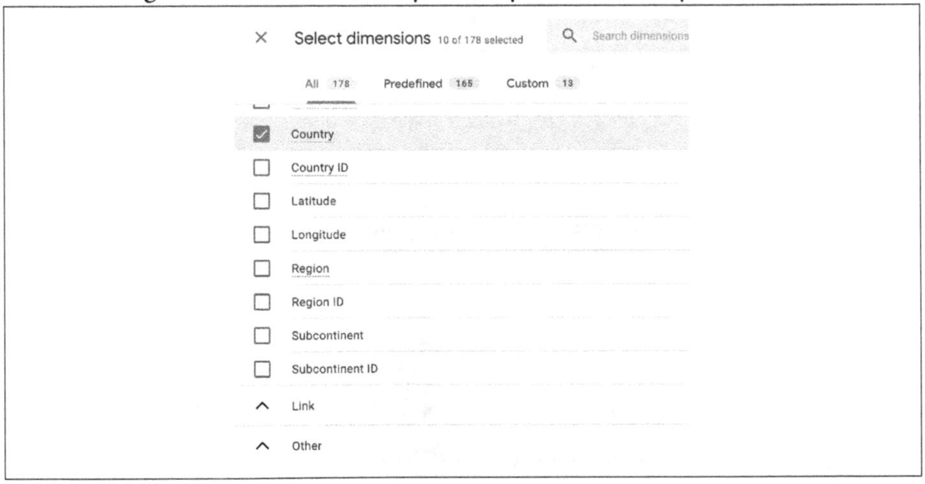

Figure 6-14. Selecting the variables you think you will need in your exploration

3. Choose Technique: Choose the analysis technique in the next tab column. These range from tables, funnel explorations, path graphs, and segment overlap plots. The techniques all have different functionalities. For instance, in Figure 6-15, right-clicking on segment overlaps allows you to further deep-dive into those users or create an Audience and subsegment from them.

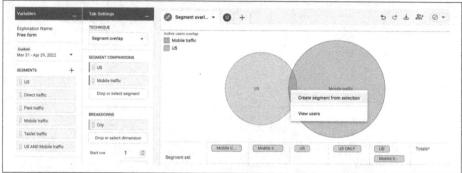

Figure 6-15. The Explore module includes various reports with different functionality; in this example, we use the Segment overlap technique to see which users are from the US and use mobile devices

4. Apply data fields: Apply segments, filters, and fields to your report. Keeping your use case in mind, bring the appropriate dimensions and metrics into the visualization technique, as shown in Figure 6-16.

Figure 6-16. Selecting the appropriate fields for your exploration report

5. Iterate and analyze: Repeat the preceding steps to bring in more fields, segments, and filters to get to the information you require. Once you're ready, you have the choice to share the data with another GA4 user or export to PDF or Google Sheets.

The meat of your analysis depends on the various Exploration techniques, which is hopefully a growing list of tools, all with right-click interactivity to help with the "flow" of your analysis. Since these will be the largest factor in extracting insight from your GA4 data, here is a brief tour of those techniques (as of the date I'm writing this) and features you may use:

Free-form exploration

This is the typical place to start since it includes a traditional table and plotting options such as line, scatter, bar, and geo reports. Using the line charts will activate time-series features such as anomaly detection to highlight when your measurement saw unusual activity. An example was shown previously in Figure 6-16.

User exploration

This report lets you drill down to individual `cookieIds` and offers a very fine level of detail, an example of which you can see in Figure 6-17. You can use this to explore users in a particular segment to see what events they triggered. From here you can also delete user data if necessary. A good use case is to segment users and find a typical example of behavior you want to target, such as people who failed to purchase after they clicked a certain internal banner. You can then find all similar users and create an Audience to target them, perhaps in an A/B test via "Google Optimize" on page 201.

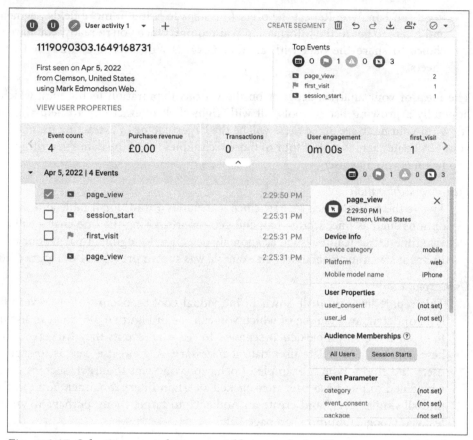

Figure 6-17. Selecting events lets you quickly create look-alike segments and Audiences for users who had similar behavior

Segment Builder

This technique lets you visualize Venn diagrams of your segments and helps you create subsegments Figure 6-15 shows an example.

Path exploration

This lets you answer questions relating to user flow, such as, "Where did users end up after viewing this page?" Because GA4 is event based, this can be extended to "What events happened after this click/view/purchase/etc.?" The "after" can be within the same session or across multiple sessions, and you can mix between event names and page titles. For instance, the banner I used as an example in Figure 6-6 has a link to my post about writing this book—are people clicking it? I can examine the flow of how many views that page got as shown in Figure 6-18.

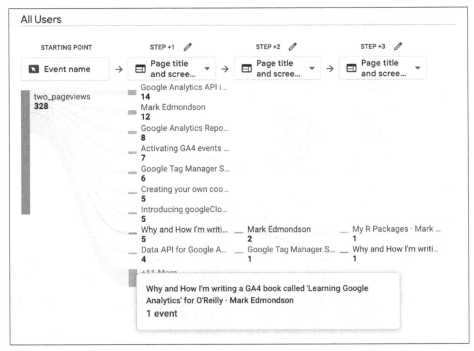

Figure 6-18. Path analysis of which pages were visited after the two_pageviews *event was triggered*

Funnel exploration

Funnels are a common technique in digital marketing where we imagine users progressing from one page to another, such as a product page to add to basket to payment page to completion. Users are assumed to enter at the top of this funnel (or starting page) and make their way in predictable steps to the end of the funnel. Concentrating on optimizing this journey to minimize users dropping off (or not progressing to the next step) is a common optimization technique to improve conversion rates. This technique is related to path analysis but focuses more on the rates and dropouts as users traverse your predetermined funnel. This is often a source for optimization, such as improving click-through or ecommerce conversion rates, and can be an important starting place to see where future data projects should focus. Funnel steps can also be events or page views. Within the funnel visualization, you can also right-click to get a User Exploration report (as seen in Figure 6-17) on who dropped out, something which was much trickier to achieve in Universal Analytics. Figure 6-19 shows an example of this.

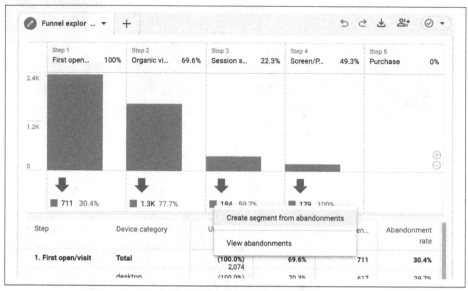

Figure 6-19. Examining a funnel dropout journey with the option to drill down into which users failed to make the next step in your funnel

Cohort exploration

Cohorts are more concerned with bucketing your users than how often they visit or come back to your website. This can help measure the "stickiness" of your website and can be a KPI if you are running a publisher website that relies on ad revenue for income. You can break down the cohorts by segment and other dimensions and decide the criteria for when a user is first counted as a visitor. For example, since I have an event firing for viewers of my Google Analytics content, I can compare that cohort to see if they come back as often as viewers of my BigQuery content (Figure 6-20).

This section covered the many visualization options available within the GA4 interface. However, you may not want to use GA4's reports since you have existing visualization tools and workflows, or you don't want to give access to GA4, or you prefer business users be in a more controlled environment. For those visualization needs, we now look beyond GA4 at other visualization tools you may use, the first being Google Data Studio.

	MONTH 0	MONTH 1	MONTH 2	MONTH 3	MONTH 4	MONTH 5	MONTH 6
Each cell is the sum of **Active users** for users who had **Any event**, in that month after googleanalytics_viewer							Based on device data only
All Users Active users	5,469	334	98	62	23	15	6
Oct 1 - Oct 31, 2021 1,052 users	1,052	71	20	23	13	8	6
Nov 1 - Nov 30, 2021 911 users	911	63	30	20	13	9	
Dec 1 - Dec 31, 2021 565 users	565	63	19	11	3		
Jan 1 - Jan 31, 2022 1,516 users	1,516	97	40	22			
Feb 1 - Feb 28, 2022 607 users	607	61	16				
Mar 1 - Mar 31, 2022 586 users	586	51					
Apr 1 - Apr 29, 2022 554 users	554						

Figure 6-20. How many users who triggered the `googleanalytics_viewer` event came back to the website the subsequent months?

Data Studio

One could argue that many less-experienced users who used to log into the Universal Analytics should now instead log into Data Studio connected to GA4. Advanced users could still use the GA4 interface for configuration and advanced analysis, but the majority of light business analytics use may be better performed with Data Studio.

Can Data Studio Do Everything?

It *is* possible to complete an entire data project using Data Studio's capabilities: you can connect it to data sources for ingestion, store the data within Data Studio tables, and do some modeling via its joins or calculated metrics. For small projects, this is by far the quickest way to get going. However, I caution against doing complicated projects solely within Data Studio. At some point, you'll be using a tool that is not optimal for the job you're trying to do with your data (e.g., modeling), and it would be better to use another tool such as BigQuery SQL to do the same job, otherwise you'll end up wasting time and resources. Rule of thumb: it's best to keep Data Studio in the role it's best suited for—visualization—and leave the transformations, joins, and so on to another.

With GA4, you have two options as a data source for Data Studio: the Data API or the BigQuery GA4 raw data export. The Data API is quicker to set up and gives you access to the same data as used in the Standard Reports, but it is harder to create advanced reports such as funnels and segmentations. BigQuery gives you access to any data you need but can involve complicated SQL to get it out.

I would argue that the best use of Data Studio is as an analysis tool, which is easy enough to use for analysts to operate on their own without needing predefined dashboards created by you. Perhaps you can get them started with templates, but I think the main purpose of Data Studio is that almost anyone spending some time should be able to get simple line chart or plot relatively quickly, if they have, say, comparable skills to using the Microsoft Office suite. To make this as easy as possible, those analysts need to be able to connect to nice clean, tidy, useful aggregated tables that I would stress should be a focus of your data engineering. Working on making those tables as useful as possible for others to build their own personalized analysis on top may bring better value than trying to make a perfect dashboard for every user.

When connecting to Data Studio, you have two options: the direct Google Analytics connector that uses the Data API or the BigQuery connector to use GA4's raw data export. Within Data Studio, click on the "Resource" menu, then "Add data to report," which will pull up the list of possible data sources shown in Figure 6-21.

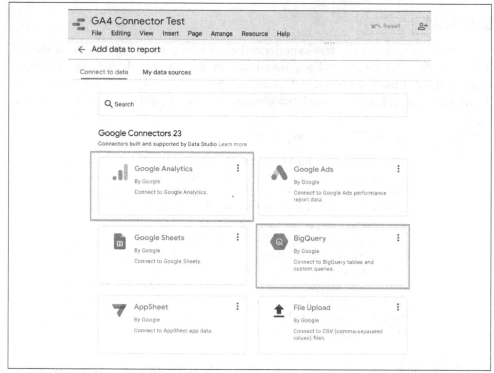

Figure 6-21. You can use the Google Analytics connector top-left to connect via the API, or if you have the GA4 BigQuery data export live (and I recommend it!), you can also connect via the BigQuery connector middle-right

Both have support to help make it easier to surface the metrics you may need and are equivalent at a top level. Just to check this, in Figure 6-22, I compare two tables of Event name and Event count and show that they give the same numbers even though one table is sourced from the Google Analytics connector and the other from BigQuery.

GA4 Connector

	Event name	Event count ▼
1.	page_view	7,269
2.	fetch_user_data	4,856
3.	user_consent	4,713
4.	article_read	4,250
5.	session_start	3,934
6.	r_viewer	3,567
7.	user_engagement	3,131
8.	CLS	2,905
9.	LCP	2,822
10.	first_visit	2,176
11.	scroll	1,853
12.	gtm_viewer	1,736
13.	googleanalytics_viewer	1,613
14.	docker_viewer	1,557
15.	FID	1,524
16.	bigquery_viewer	1,039
17.	click	595
18.	two_pageviews	339
19.	optimize_personalization_impression	116
20.	r_package_loaded	111

1 - 20 / 20 ‹ ›

GA4 BigQuery

	Event Name	Event Count ▼
1.	page_view	7,269
2.	fetch_user_data	4,847
3.	user_consent	4,714
4.	article_read	4,250
5.	session_start	3,934
6.	r_viewer	3,567
7.	user_engagement	3,132
8.	CLS	2,906
9.	LCP	2,823
10.	first_visit	2,176
11.	scroll	1,853
12.	gtm_viewer	1,736
13.	googleanalytics_viewer	1,613
14.	docker_viewer	1,557
15.	FID	1,526
16.	bigquery_viewer	1,039
17.	two_pageviews	809
18.	click	595
19.	optimize_personalization_impression	116
20.	r_package_loaded	111

1 - 20 / 20 ‹ ›

Figure 6-22. Connecting Data Studio via the Google Analytics connector versus the Big-Query table

Data Studio is popular, so there are many templates already in the Data Studio gallery for GA4 that you can use or base your own designs on. As a random example, the dashboard from Data Bloo shown in Figure 6-23 lets you switch between GA4 and Universal Analytics and shows data from the example Google Merchandise Store.

Figure 6-23. A GA4 Data Studio template example for GA4 from Data Bloo

Data Studio is the best option for many people because it's free, powerful, and has a lot of native integration with GA4 and other connectors both inside and outside the Google suite. However, you may start to feel its boundaries if you're looking for more complicated data transformations, user management, or interactivity with other data services. In that case, you may need a solution that takes care of the data processing funnel; Looker is one option, which we'll talk about in the next section.

Looker

Looker is much more than a visualization tool—it's what is known as a more general business intelligence (BI) tool. Looker takes care of data definitions across all of your datasets and tries to unite them under a single source of truth with your own business logic on top. It comes with its own SQL-like language, LookML, to achieve this. The idea is that your data engineers work with LookML to unite your datasets, tidy that data up, and establish a naming convention that can then be exposed downstream via Looker applications including apps, visualizations, Data Studio imports, or elsewhere.

Looker does a lot more heavy lifting in data activation terms since it can turn your raw data or models into datasets ready for your business users. Looker "looks" into your existing datasets, such as on BigQuery, and can combine them with other services, even if those services lie with other cloud providers outside of Google or on your own on-premises databases. Looker executes SQL on your behalf with those services and adds a centralized place for all your business logic. The SQL Looker executes doesn't need to be exposed to your end users, so much so that they can perform complicated queries like aggregations and joins across several datasets via Looker's drag-and-drop interface. However, this all comes at a cost, so Looker should be regarded as an enterprise tool compared to Data Studio.

There are integrations between Data Studio and Looker that let you connect Data Studio to Looker datasets after your business rules have been applied. This is helpful since you can maintain the data governance that Looker provides but also let users analyze that data by themselves and combine it with ungoverned data easily via Data Studio. This lets you have the best of both worlds by keeping data analysis democratic and easy to use via self-service Data Studio, but still keeping standards on your data to avoid incorrect conclusions that may affect your business performance.

Looker has an existing integration with GA4 that mirrors a lot of the functionality Universal Analytics had and more. Since Looker offers the LookML business logic language, it can be used to engineer the sometimes complicated SQL necessary to create funnels, sessions, and trends out of a raw BigQuery GA4 export. You can see some of the reports it can create in Figure 6-24 and read more about Looker's integration with GA4 via the Looker Marketplace (*https://oreil.ly/1aqNQ*).

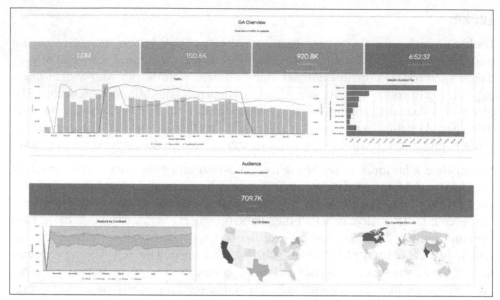

Figure 6-24. Looker connects to GA4's BigQuery dataset and uses its LookML language to create useful data points

Even if you're not using Looker's visualizations, it may be beneficial to connect it to its preexisting modeling via "Looker blocks"—a detailed overview of which can be found on Looker's GitHub profile (*https://oreil.ly/XD9rk*).

This will create aggregated tables out of your raw GA4 export and may save you a lot of work re-creating the same yourself. As an example, it calculates landing/exit pages and what digital channel a user's session belongs to, and utilizes BigQuery ML (see "BigQuery ML" on page 177) to create some purchase propensity modeling. Bear in mind, though, that it does create additional tables that could effectively double the size of your GA4 data, with associated doubling of costs.

If you're using Looker, you will likely have other datasets that you're connecting to as well, so it may be useful to join those datasets on GA4's userId or another custom field.

Google owns Data Studio and Looker, but they are by no means the only options you have for visualization tools, as we shall see in the next section.

Other Third-Party Visualization Tools

There are many other non-Google-based visualization tools out there, which you could use if you prefer. If I were looking at another visualization tool, I would consider the following points:

- Does the dashboard connect via the Data API, BigQuery GA4 export, or manual uploads of data? I would prefer the Data API for simple or real-time reports and BigQuery exports for more complicated reports such as funneling, but keep in mind that the technical cost of creating the SQL to model those reports may not be trivial. I see little reason to rely on manual exports of data.

- Are you already using the visualization tool for other business functions? This can be a compelling reason to import into an existing tool everyone is already trained in rather than insisting on yet another new tool.

- Does your visualization tool cover Google user management and report distribution well?

Of all the visualizations tools out there, the most common third-party tools are Tableau and Power BI. Both are solid choices, and I would say if you're already using them for other visualization tasks on your existing data stack, then stick with them; however, if you're looking to start afresh, then having everything on the Google stack is easiest, as the integrations between services is tight (e.g., you can explore BigQuery tables in Data Studio from the BigQuery interface).

Whichever visualization tool you are using, having nice and clean datasets to connect to will make things 10 times easier for your users, which we discuss in the next section.

Aggregate Tables Bring Data-Driven Decisions

As described in Chapter 4, having clean aggregated tables for your dashboard users to connect to will enable more self-service analysis for your business and is a true step toward becoming a data-driven business. I suggest that creating easy-to-use data sources above and beyond the raw data exports should be among your first steps when you're making data more of an influence in your everyday work. Making the process of analysis as easy as possible with a low entry barrier is one of the dreams that data visualization tools offer, but that dream is only as good as the quality of data your employees work with.

The more you work with your raw data to create those clean tables, perhaps using some of the techniques described in Chapter 5, the less of a bottleneck your data insight funnel will be. In my mind, a successful data visualization stream for data activation would involve the following:

- An ever-expanding list of use cases that can be fulfilled by existing and new aggregated, tidied, joined, and filtered tables ready for consumption.

- An analysis and visualization tool that the majority of your staff are trained enough in that they can self-service bespoke analysis. Aim to achieve a minimum level of competence by all members so key analysis members do not become bottlenecks.

- Regular cross-department meetings discussing data needs, and a searchable archive of existing visualizations, datasets available, and outcomes achieved.

- A special core of advanced users creating more general visualizations suitable for a wider audience.

- Data that is trusted and has a transparent source of truth. Incorporate regular QA, monitoring, and checks to avoid data mistakes, and communication when pipelines go down.

- Data visualizations from your tool that regularly used within internal communications—perhaps even links to specific views of dashboards to answer queries across the company.

Successful data visualization requires good management of the data before it hits the data visualization tool. Included in the shape and size of the data users get to work with, you should also consider infrastructure concerns such as data caching and cost.

Caching and Cost Management

Linked concepts are data caching and the costs associated with data visualization. Hopefully, if successful, your dashboards will generate a lot of calls to your data warehouse. A lot of these tools will cache the results, meaning that a repeated call for the same information will not result in a charged call to your database but will be read from that cache instead, with speed and cost advantages. However, this may not be the case for all dashboards—for instance, real-time dashboards will always require new information and should not cache data.

Related to this, you should carefully consider the type of table you're calling from the data visualization tool. For example, calling a View that is doing a SELECT * across all your columns in a table will be very expensive if carried out several times per day per user. This cost could be almost eliminated if instead the dashboard is linked to a table that is generated each morning with the same data. Some of this work will involve creating that data pipeline to create the table, as discussed in Chapter 4, and some tools may have configuration options to help. For instance, BigQuery has its BI Engine (*https://oreil.ly/5QoCA*) that gives you a cache to help with just these types of queries, or consider using materialized views, which incrementally add new data.[1]

We have read about the two most common data activations so far, creating Audiences for export within Google Marketing Suite products and creating visualizations, but probably the most impactful (yet less-common) data activation is sending data via

---

1 You can read more about materialized views in the BigQuery documentation (*https://oreil.ly/ypkiy*).

APIs to various services that can both enhance the methods described so far and unlock potentially more applications. We look at how to create these in the next section.

Creating Marketing APIs

Marketing API is a term I'm using for data activation to cover making your data available for consumption by programmatic code that falls outside of the methods described so far (i.e., visualizations and GA4 Audiences). Under the hood, both of these methods use APIs to transfer data between services, so we're looking to go one level deeper to have more control over what data can be sent and to where. APIs are a standard way of transferring data across a broad range of programming applications, in any coding language and for many different types of apps. We're looking at creating API endpoints that respond to requests for data with typically a JSON data packet response, much like if you were calling GA4's Data API, but customized to your own business.

Creating Microservices

There are many tools within both Google Marketing Suite and Google Cloud that can help you make marketing APIs and let you create, scale, and monitor them easily. With these tools, you can start to create bespoke data services that target specific data applications of interest to your business, such as returning the number of subscriptions a user has if you send your API a `userId`. These are commonly named *microservices* since you can have many independent services available, which makes it easier to mix and match what you need.

One framing of GTM SS could be to call it a digital marketing API development kit that you use as a platform for creating microservices. The clients within its interface effectively create URL endpoints for you to use, and then its WebUI provides the control mechanism for how they trigger and what data they process. Finally, the templates and tags within GTM SS let you send that data on. At first pass, this is usually GA4 related, but there's nothing stopping you from creating API endpoints for your own microservices. For example, GTM SS can connect to Firestore, where your user information can reside. You can create a client, trigger, and tag that create a microservice that returns user information when you send a user ID to your URL endpoint, such as `/user-info?userid=12345`. An advantage of using GTM SS over other systems is that you can more easily apply the same level of control as you have with your web analytics data streams and will be in a familiar interface for your digital marketers who are using that data.

Google Cloud has several other services for creating APIs:

Cloud Functions

Cloud Functions can be called via HTTP triggers, which effectively means you can run code in response to an HTTP call, compute, and send back data. This is often the easiest way to get going since you need only upload some code off its supported languages, hit publish, and you're done.

Cloud Run

Cloud Run is more flexible but needs a bit more work than Cloud Functions because it runs Docker containers. This means it can run almost any code and environment, unlike Cloud Functions, which only run on supported languages.

App Engine

App Engine is a step more complicated than Cloud Functions or Cloud Run but gives you more control over what server resources are dedicated to your code. If you want more control over costs and autoscaling, it may be a better option. App Engine also has more integrations with other GCP services because it's been around for much longer.

Cloud Endpoints

This doesn't run your code but is a proxy in front of your API, which is helpful when you need common API management functions such as authentication, API keys, monitoring, or logging.

Firestore

When populating data into your API, you'll most commonly do so by fetching data from a Firestore instance rather than, say, BigQuery. This is because Firestore will be much more performant in returning data quickly.

Microservices can be the secret to upgrading your digital analytics tech stack to perform better. Once you have them in place, they are very reusable due to their independent nature and can scale up over many use cases. Examples of my use of microserves in the past include outputting forecasts of search trend data, predicting whether a campaign will hit targets, and returning which audience segment a particular user belongs too. And because they run over the universal HTTP standard, you can call them from any language or even via spreadsheets.

The activations we've talked about to date have relied on mostly reading data, but how can we get our data to react to incoming data? For this, we need to consider event-based triggers, for which GA4 is well suited given its new data model, as shown in the next section.

Event Triggers

Since GA4 uses events for its measurement system, it can take particular advantage of triggers that have applications beyond only measurement. You can, for instance, fire events based on page views, normal click events, purchases, and actions, and also if a user enters an Audience, as shown in Figure 6-2 earlier in the chapter. This opens up powerful data activation techniques, since that user's data can be sent to many different activation platforms beyond just those on the Google Marketing Suite.

The next section provides an example of how you can implement this.

Streaming GA4 events into Pub/Sub with GTM SS

For this example, we'll send a `send_email` event from a website to GTM SS and then send that event on to Pub/Sub, an event messaging system we talked about in "Pub/Sub" on page 114. We use Pub/Sub because it's detached from any specific service, so you can quickly adapt it for your own use cases by pointing it at a different application that will react to its message.

We first need a tag that will send your GTM events on to an HTTP service. The code in Example 6-1 is generic enough that you can supply any URL you control.

Example 6-1. GTM SS tag code to turn a GTM event into an HTTP request. The code example is simplified; for production, you may wish to extend the logging information and/or add a private key to the HTTP request.

```
const getAllEventData = require('getAllEventData');
const log = require("logToConsole");
const JSON = require("JSON");
const sendHttpRequest = require('sendHttpRequest');

log(JSON.stringify(data));

const postBody = JSON.stringify(getAllEventData());

log('postBody parsed to:', postBody);

const url = data.endpoint + '/' + data.topic_path;

log('Sending event data to:' + url);

const options = {method: 'POST',
                  headers: {'Content-Type':'application/json'}};

// Sends a POST request
sendHttpRequest(url, (statusCode) => {
  if (statusCode >= 200 && statusCode < 300) {
    data.gtmOnSuccess();
  } else {
```

```
        data.gtmOnFailure();
    }
}, options, postBody);
```

The GTM tag calls for two data fields to be added:

data.endpoint
> This will be the URL of your deployed Cloud Function given to you after your deployment, which will look something like *https://europe-west3-project-id.cloudfunctions.net/http-to-pubsub*.

data.topic_path
> This is the name of the Pub/Sub topic it will create.

Once implemented, you should be able to create a tag from the template that will look a little like Figure 6-25. The screenshot shows that this tag has been set up to trigger on a `form_submit_trigger`, but this trigger can be anything you wish, following the normal rules of GTM.

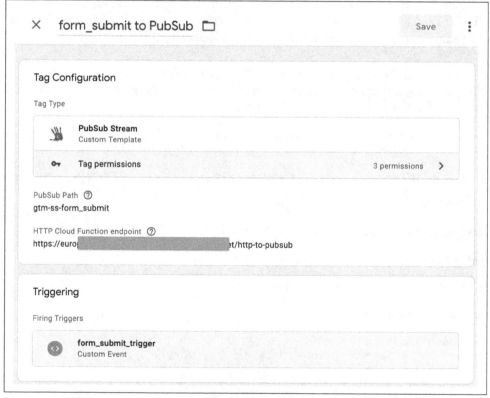

Figure 6-25. The tag within GTM SS for forwarding on your events to an HTTP endpoint

The URL you want to forward to can be a Cloud Function that runs the code shown in Example 6-2.

Example 6-2. Python code within a Cloud Function for receiving the HTTP request with your GA4 data and sending it onward to a Pub/Sub topic. The print() logs may represent a significant cost at high data volumes, which you may choose to remove.

```python
import os, json
from google.cloud import pubsub_v1 # google-cloud-pubsub==2.8.0

def http_to_pubsub(request):
    request_json = request.get_json()

    print('Request json: {}'.format(request_json))

    if request_json:
        res = trigger(json.dumps(request_json).encode('utf-8'), request.path)
        return res
    else:
        return 'No data found', 204


def trigger(data, topic_name):
  publisher = pubsub_v1.PublisherClient()

  topic_name = 'projects/{project_id}/topics{topic}'.format(
    project_id=os.getenv('GCP_PROJECT'),
    topic=topic_name,
  )

  print ('Publishing message to topic {}'.format(topic_name))

  # create topic if necessary
  try:
    future = publisher.publish(topic_name, data)
    future_return = future.result()
    print('Published message {}'.format(future_return))

    return future_return

  except Exception as e:
    print('Topic {} does not exist? Attempting to create it'.format(topic_name))
    print('Error: {}'.format(e))

    publisher.create_topic(name=topic_name)
    print ('Topic created ' + topic_name)

    return 'Topic Created', 201
```

Deploy the code via the following:

```
gcloud functions deploy http-to-pubsub \
        --entry-point=http_to_pubsub \
        --runtime=python37 \
        --region=europe-west3 \
        --trigger-http \
        --allow-unauthenticated
```

Once deployed, you should see a generated URL you can put in the GTM SS trigger we saw in Figure 6-25.

With these two generic code deployments, you'll have your choice of GA4 events streaming into a Pub/Sub topic to do with as you wish: this is powerful! The scripts take care of pushing data to where you need it, but if you want your events to also read data, then I recommend using Firestore, which is covered in the next section.

Firestore Integrations

Firestore is ideally suited to as a backend for your marketing APIs because it will have high performance in returning data when you have a key to save that data under. The nature of Firestore is that you typically supply an ID, and then any data underneath that ID can be returned.

How you get the data into Firestore will depend on its source. You may be importing user ID–level data from your CRM systems. In that case, you'll need to set up scheduled imports to populate your Firestore database. Other data imports such as product information may lie in different systems that a data pipeline will need to be created for.

An example of data you could be populating is shown in Figure 6-26.

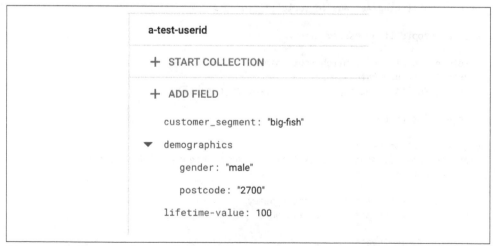

Figure 6-26. Example data within a Firestore instance

The application code for your marketing API would then typically need to handle the following:

1. Receive an HTTP call at your endpoint with a `userId` included, e.g., `https://myendpoint.com/getdata?userid=a-test-userid`.

2. Parse out the `userId` and use it to fetch a document from your Firestore. In Python, you would do this with `doc_ref = db.collection(u'my-crm-data').document(u'a-test-user-id').get()`.

3. Return the Firestore data in your HTTP response body.

Within GTM SS, you have a smooth integration with Firestore via its Firestore Lookup variable, found in the list of default variables available within GTM SS. You can use this variable to directly insert Firestore documents into your tags and clients. The GTM SS template API also supports writes to a Firestore database, as shown in Figure 6-27.

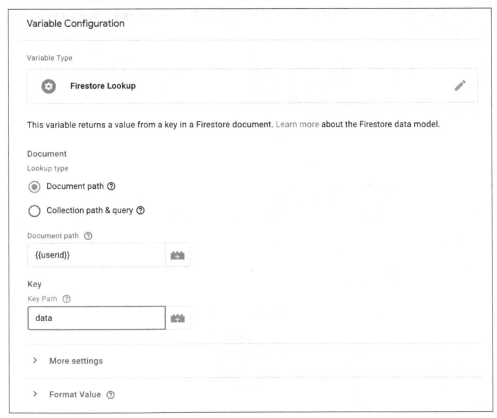

Figure 6-27. A Firestore Lookup variable in GTM SS

Adding Firestore to your marketing portfolio enables a lot of cool applications, and GTM SS gives you a familiar gateway to utilize them. You will likely have data within BigQuery that you also want to appear in Firestore—the next section looks at techniques you can use to achieve this data pipeline.

Importing BigQuery into Firestore

We've talked a lot about BigQuery for its data analysis and modeling abilities and Firestore for its speed of data access and ability to work with structureless nested data. A common need is to export the results of your BigQuery table to Firestore.

Firestore and BigQuery have different approaches to data storage, but you can create an export if you can choose which column within your BigQuery table should be the key for Firestore: a `userId` is the usual choice. The solution shown in Example 6-3 offers a Cloud Composer DAG that triggers a Dataflow job. You don't need to know the details of the Dataflow job since it's all self-contained within Docker; instead, we pass in the appropriate column we would like the Firestore key to map to from the BigQuery column and pass it the BigQuery table we have most likely created in a previous step.[2]

Example 6-3. A Cloud Composer DAG to create a BigQuery table and send it to Firestore via Dataflow. In this case, the BigQuery SQL to create the BigQuery table with one column containing a userId is assumed to be in a file named ./create_segment_table.sql.

```python
import datetime
from airflow import DAG
from airflow.utils.dates import days_ago
from airflow.contrib.operators.bigquery_operator import BigQueryOperator
from airflow.contrib.operators.gcp_container_operator import GKEPodOperator

default_args = {
    'start_date': days_ago(1),
    'email_on_failure': False,
    'email_on_retry': False,
    'email': 'my@email.com',
    # If a task fails, retry it once after waiting at least 5 minutes
    'retries': 0,
    'execution_timeout': datetime.timedelta(minutes=240),
    'retry_delay': datetime.timedelta(minutes=1),
    'project_id': 'your-project'
}
```

---

2 The original Dataflow code was created by Yu Ishikawa, which you can see on his GitHub profile (*https://oreil.ly/GhMc4*).

```
PROJECTID='your-project'
DATASETID='api_tests'
SOURCE_TABLEID='your-crm-data'
DESTINATION_TABLEID='your-firestore-data'
TEMP_BUCKET='gs://my-bucket/bq_to_ds/'

dag = DAG('bq-to-ds-data-name),
          default_args=default_args,
          schedule_interval='30 07 * * *')

# in production sql should filter to date partition too e.g. {{ ds_nodash }}
create_segment_table = BigQueryOperator(
    task_id='create_segment_table',
    use_legacy_sql=False,
    write_disposition="WRITE_TRUNCATE",
    create_disposition='CREATE_IF_NEEDED',
    allow_large_results=True,
    destination_dataset_table='{}.{}.{}'.format(PROJECTID,
                                                DATASETID, DESTINATION_TABLEID),
    sql='./create_segment_table.sql',
    params={
        'project_id': PROJECTID,
        'dataset_id': DATASETID,
        'table_id': SOURCE_TABLEID
    },
    dag=dag
)

submit_bq_to_ds_job = GKEPodOperator(
    task_id='submit_bq_to_ds_job',
    name='bq-to-ds',
    image='gcr.io/your-project/data-activation',
    arguments=['--project=%s' % PROJECTID,
               '--inputBigQueryDataset=%s' % DATASETID,
               '--inputBigQueryTable=%s' % DESTINATION_TABLEID,
               '--keyColumn=%s' % 'userId', # must be in BigQuery ids (case sensitive)
               '--outputDatastoreNamespace=%s' % DESTINATION_TABLEID,
               '--outputDatastoreKind=DataActivation',
               '--tempLocation=%s' % TEMP_BUCKET,
               '--gcpTempLocation=%s' % TEMP_BUCKET,
               '--runner=DataflowRunner',
               '--numWorkers=1'],
    dag=dag
)

create_segment_table >> submit_bq_to_ds_job
```
```

There are other ways to import from BigQuery to Firestore that you may prefer since
the preceding example will only work on a paid Airflow server such as Cloud Com-
poser. Krisjan Oldekamp shows an approach using Google Workflows that you can
```

read about on the Stacktonic site (*https://oreil.ly/gnG12*) that works for smaller data. I'm sure more direct methods will also surface since this is such a useful direction for a digital marketing pipeline to take.

Summary

In this chapter, we looked through several ways to activate your GA4 data once you have collected, stored, and modeled it. GA4 has many features built into it that can give you quick results: the Audience features that can export to various other products within its Google Marketing Suite and the visualization and analysis tools within GA4's WebUI. However, it doesn't stop there because its event structure makes integrations with other systems easier than ever before by allowing you to send your data either via BigQuery or real-time streams via GTM. This opens up virtually any other data activation product out there. The main lesson from this chapter is to consider data activation front and center when you're formulating your use case and to not regard it as an afterthought. Getting it right will give you visible results that you will be able to show off to your colleagues and will help you win more budgets for future projects.

The chapters so far have been a lot of theory on what you may need, but you can only truly appreciate them if they're put to practical use. In the next chapter, we begin looking at the use case examples, putting into practice some of the techniques we've talked about in the previous chapters.

## CHAPTER 7

Use Case: Predictive Purchases

This chapter's use case serves as a simple example to accustom you to the structure shared with the more complex use cases we'll see in later chapters. We'll use only one platform, GA4 , to create it. However, the same data roles apply for the more involved use cases later, and we show that it's possible to swap out those data roles should doing so better serve your needs.

In this scenario, let's suppose that you're a book publisher who would like to advertise your incredible new guide to Google Analytics. You have a custom GA4 set-up where the categories of books that your customers are browsing are recorded and purchase behavior from thousands of transactions are available. You're also running a lot of Google Ads search campaigns tailored for each category, but because the subjects you are marketing against are broad, you're catching a lot of impressions for customers just searching for general information. So you're spending money on campaigns that don't target potential customers and therefore cost you more money than you would like. You also have a theory that you're potentially spending money advertising to people who would purchase the book anyway and would like to see if you can suppress advertising for those customers so you can spend more budget on impressions for customers who may need to be persuaded that this really is the book for them. Using GA4, we're going to set up an Audience for those users with a 90%+ probability of purchasing and suppress advertising for them with the hope that it will make the campaign more efficient in total.

With hope in your heart, you approach your boss to see if they will approve the resources to put such a plan into action. You're looking for a quick win with minimal resources required so you can get approval for your even more ambitious future plans. Your boss will ask for the business case, which we'll cover in the next section.

Creating the Business Case

Predictive purchasing uses modeling to predict whether a user will buy in the future. This can be used to alter site content or advertising strategy for those users. For instance, if a user is certain to purchase with a probability above 90%, we want to suppress marketing to that user because the job of persuading them to buy is already done. Conversely, if a user is predicted to churn in the next seven days, perhaps we can give up on them as a lost cause. Enacting such a policy means that you can move your budget allocation to target only the users who may or may not buy. This should increase your ROI and lift your sales revenue. This is a generic description, but for your business case, you will need estimates of the real numbers involved. This assessment of value is a key first step in showing the value of your use case.

Assessing Value

We first look at the hypothetical revenue from Google Ads campaigns that we wish to enable predictive conversions for; the current standings are given in Example 7-1.

Example 7-1. Accountancy numbers of Google Ads for this use case

```
Google Ads monthly budget: $10,000
Cost per click: $0.50
Clicks per month: 20,000
Conversion rate: 10%
Average order value: $500
Orders: 2000

Monthly Revenue: 2000 orders * $500 = $1,000,000
Monthly ROI: $1,000,000 / $10.000 = 100
```

We propose that we'll get the same conversions for users whose probability to convert is over 90% if we don't advertise to them. Our assumption is that advertising is not changing these user's behavior once they have made a decision to buy. This is an assumption we should confirm when looking at the results of the project.

We would like to focus our $10,000 budget at the remaining users who have less than a 90% chance of converting. We estimate that being able to bid higher for these users will result in a 10% uplift in clicks for them.

Assuming our conversion and average order value stays the same, this should result in an uplift of $90,000 a month in revenue for the same Google Ads cost, giving us a monthly ROI of 109, as shown in Example 7-2.

Example 7-2. Accountancy numbers when the use case is enabled

```
Google Ads monthly budget: $10,000
Cost per click: $0.50
Clicks per month: 20,000
Conversion rate: 10%
Average order value: $500
Orders: 200 (top 10% who buy anyway) + (1800 * 10% uplift) = 2180

Monthly Revenue: 2180 orders * $500 = $1,090,000
Monthly ROI: $1,090,000 / $10.000 = 109

Expected uplift: $90,000 a month
```

This gives us a value for the amount of uplift cost we hope to achieve, which we now need to subtract from the cost of doing it to see if it's worthwhile.

Estimating Resources

As we are using native integrations within GA4, the total resources needed will be minimal. We'll need configuration time to configure GA4 and the Audience exports to Google Ads, but no third-party services will be needed, and there are no GCP running costs. This is a big advantage to running native GA4 integrations if they fit your use case well.

Your GA4 will need to be configured for ecommerce tracking, and consent choices should be captured to avoid targeting users who have not opted in. We'll assume this has already been done via your initial analytics implementation of GA4. The work to create the Predictive Audience and export it to Google Ads should be well in scope for your existing digital analysts, who are used to working with similar projects.

Data Architecture

The data flows are basic here since we need only GA4 and Google Ads—see Figure 7-1. The diagram is simple, but this will quickly become more complicated when more data sources are involved.

Use cases that take advantage of the native integrations within GA4 are most likely the lowest hanging fruit for your analytics implementations. A good baseline is to enable all of these if possible, which will put you in good shape for extending to more advanced use cases later. This base level involves matching Audiences for all Google Marketing Suite services, such as Google Ads, Google Optimize, and Search Ads 360.

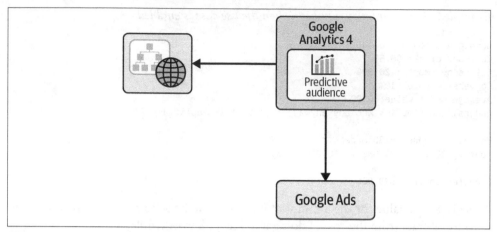

Figure 7-1. Data architecture for the Predictive Audiences use case: website data is sent to GA4, which creates the predictive audience that is then exported to Google Ads

To cover this use case, we'll first check that your GA4 streams have filled the data collection requirements.

Data Ingestion: GA4 Configuration

This section looks at what data you'll need to enabled for the use case to be successful, with considerations as outlined in Chapter 3. For this simple case, we only need to worry about GA4 configuration. We'll cover some extra requirements related to Predictive Audiences in the paragraphs that follow.

To enable Predictive Audiences, you need a predictive metric to work on your GA4 data streams. Google help files explain (*https://oreil.ly/JCBRA*) how to enable these.

The most important requirements are listed here:

- You need enough purchasers to trigger the model. For this, you need at least 1,000 conversions and at least 1,000 visitors who did not convert in the last seven days, and those users must be returning visitors.

- You need to send ecommerce Recommend Events (see "Recommended events" on page 51), including the `purchase` event or `in_app_purchase` for mobile apps.

- You need to enable the data-sharing benchmark setting within GA4 so that the model can benefit from shared aggregate and anonymous data from other properties (and, conversely, they will benefit from your anonymous data).

- You need to use as many GA-recommended events as possible within your GA4 property because these may feature in the model to improve accuracy.

- You need to connect Google Ads to the GA4 account, which is necessary to export the Audience for use with your Google Ads team and to enable personalized advertising.

- You need to enable Google Signals (see "Google Signals" on page 68) in your GA4 settings to link user data between GA4 and Google Ads.

 Since you're working with targeting data, you should also consider user privacy. You may also need to gather consent from users that their data can be used for re-marketing campaigns, an extra consent on top of statistical usage. In that case, to build your Audience, you will need the predictive metric and a way to distinguish which users consented to having their data exported to Google Ads.

You can do this by setting up a `user_property` that tracks a user's current consent choices. See "User Properties" on page 61 for instructions on how to set this up.

Once done, the `user_consent` and `event_consent` dimensions should be available to qualify your Audiences.

You may need to check or enable the data collection requirements, but if you have a standard ecommerce implementation and enough volume, you may already be covered. If that's the case, well done! Move onto the next data role, data storage. If not, you will need to scope out a configuration project to enable the data collection. Note that a side effect of this will be to mature your GA4 data collection for use with many other use cases, so this demonstrates that a use case–led approach doesn't mean that only one business case will benefit each time. As you work your way through more and more use cases, you will find that more and more often the requirements are already covered, and you can check them off quickly and move to your other data roles.

Data Storage and Privacy Design

We're now looking at the considerations outlined in Chapter 4.

Data storage for this use case is within GA4 or with exports via Google Marketing Suite. This is another big advantage to using native integrations! Events will also be available in the BigQuery GA4 export and Data API if you need them for other applications. However, even if you're only using GA4 defaults, you may need to consider your user privacy.

Although you won't be storing data within your own systems, it's still important to be mindful of the user data you're sending to Google Analytics. Since the data required is returning users, cookies will be a factor, so you will need at least a cookie consent in some regions. The data sent will be of a pseudonymous nature linked to a cookieId. Recent rulings in Europe may also require you to make sure that you're not sending IP addresses or other identifiable data to the US for European citizens, whereas US legislation may require other privacy requirements, such as ensuring you have consent. Data privacy is a constantly evolving issue so make sure you consider it within your use case design if you want to avoid legal risks in the future.

Since the application includes targeting, you may need a marketing consent to process the data. To ensure your audience has given this consent, you need to capture this consent in a user event and include it within your Audiences.

If you're turning on benchmarking to improve your models, anonymous data is shared, but you may also want to consider if this is covered by your legal permissions on cross-region data from, say, the EU to US.

If you're comfortable with your user privacy design, you can now move on to sending your user data to the data modeling phase.

Data Modeling—Exporting Audiences to Google Ads

Let's now consider the data processes as described in Chapter 5.

The data modeling in this example is all handled by GA4's Predictive Measures feature and is largely opaque in that you can do little to influence its predictions. What you gain in convenience you lose in configuration.

If you have more custom needs, such as using different data sources or output to another platform, you will need to start swapping out this role—the data modeling is usually the first place you start looking at improving beyond the defaults.

Have a Debug GA4 Account

Since the predictive measures won't alter anything before you send them to Google Ads, you can test how they look in your main GA4 property. For other use cases, you may need to send the data to a debug GA4 property first, and it's a good idea to have one available for this purpose for testing and quality assurance.

For this use case, we'll use the existing Predictive Audiences feature in GA4, which does have a small amount of configuration available.

Once you're eligible for Predictive Audiences, they will start to appear in your Audiences menu. Selecting an audience will take you to a configuration screen where you can add your extra criteria, such as the user consent status—see Figure 7-2. We're starting from a base of using the "Likely seven-day purchases" Audience.

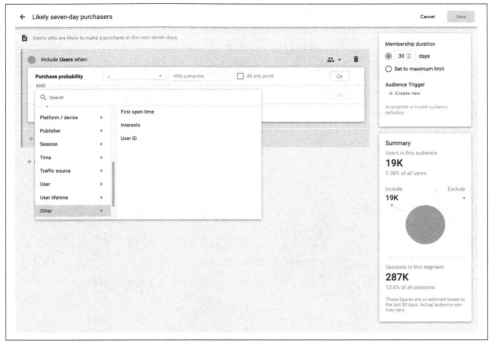

Figure 7-2. Customizing a Predictive Audience

Clicking through to configure the prediction, you can set the thresholds, which you can take from your use case design. You may want to create a few Audiences with different thresholds to experiment with to see which has the best results, e.g., 80%+ versus 95%+. You get an estimated total of how many users it would have affected if it had been live in the last 28 days, which can be helpful to judge effectiveness. The example in Figure 7-3 shows that around 32,000 purchasers would be within the group you wish to suppress advertising to.

You may also want to create an event for when users enter your segment, so you can react in other systems. For instance, in BigQuery, you will be able to segment all users who are likely to purchase in the next seven days, export that list and link it to your CRM system, and email them with, for example, a loyalty program.

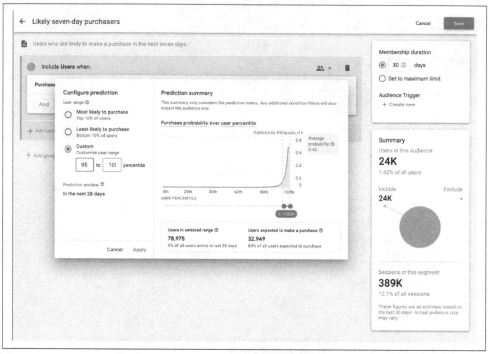

Figure 7-3. Configuration of an Audience showing likely purchasers in the next seven days

You can now create the Audience that will be exported to all your connected Google Ads accounts within 24 hours, and are ready to be activated.

Data Activation: Testing Performance

We now move to the final data activation part of the project, which was covered in Chapter 6. At this stage, you can hand the project over to the Google Ads team, who were hopefully also part of scoping the project, helping to define the Audiences they would find most helpful in execution.

Once the Predictive Audiences are exported to Google Ads, any Google Ads specialist should be able to use them to create the relevant campaigns and use those Audiences (or data segments, as they may be referred to now in Google Ads).

In your pilot stage, it's a good idea to A/B test your campaigns so you can measure if there is any relative uplift. There are a few assumptions in the use case about customer behavior when customers are exposed to ads that may not be valid.

If there is no effect, it isn't a total failure! It's valuable information about how your customers are interacting with your website—for instance, it may prove that the top 10% of your converters really rely on your advertising for that final push to sale.

Look to report back the impact on your business metric, in this case, ROI—did it hit the target you were expecting? If not, what assumptions in your set target weren't valid?

Let's compare this with our expectations from before the project in Example 7-3. We see that the assumption that the top 10% would still buy wasn't totally accurate as we had a 10% drop-off, and we got only a 5% uplift using predictive metrics to target budget at the "90% and below" segment instead of the 10% we expected. This moved our actual uplift down from $90,000 a month to $35,000 a month.

Example 7-3. Accountancy numbers when the use case is enabled (actual)

```
Google Ads monthly budget: $10,000
Cost per click: $0.50
Clicks per month: 20,000
Conversion rate: 10%
Average order value: $500
Orders: 180 (top 9% who buy anyway) + (1800*5% uplift) = 2070

Monthly Revenue: 2070 orders * $500 = $1,035,000
Monthly ROI: $1,035,000 / $10.000 = 103.5

Actual uplift: $35,000 a month
```

Generating only $35,000 extra income a month rather than the anticipated $90,000 may be regarded as a disaster, but the fact that you now have real numbers on your approaches gives you a distinctive advantage over competitors who haven't done the same, or even worse, are doing the same but not measuring it correctly. Now, for your future use cases, you have the experience to enable you to make better estimations and home in on which use cases are worth pursuing in the future. It's also likely that you will discover unanticipated side effects to your activity. Perhaps the Predictive Audience segments you created are proving helpful for your web developers to A/B test website copy aimed at loyal users.

One conclusion from your investigation may be that you need to alter the percentage bands for your predictive exclusions, for example, 50% to 99% may be a better range. You may also conclude that the data you're currently getting from GA4 is not enough to make accurate predictions, and you need to start building a business case for importing additional data into your model. The main point here is that you have learned something that you can build on for the future, and that represents a competitive advantage over other businesses who did not run the same experiment.

Summary

Predictive audiences and metrics are available only once you hit certain thresholds and with the data you have in GA4. You may be able to improve the modeling by including more data points, such as first-party data. We cover this in the next use case chapter, Chapter 8.

It may be that you want to have more control over the modeling part of the process. In this case, you will need to start considering exporting the data and creating your own model (in BigQueryML or otherwise), but when creating your own modeling, you can use the same data capture and activation from this use case. For this, the follow-up use case scope will need to include how you model and store your data but should be much more informed than if you started from scratch.

Using your own model may also be better from a privacy perspective, for example, if you wanted to use more first-party data in your model that should stay in the EU region. Within Google Cloud, you can specify where your data is processed, which you cannot do in GA4, so for that reason it may be better to process your model data locally.

Another alternative track is to change the data activation method. You may want to export your Audience to Google Optimize instead and change content on your website for those users who are predicted to buy versus those who will not. In that case, you could keep all the same data collection, storage, and modeling processes and duplicate the Audience into another third-party service, offering a consistent message for those users both on and off your website.

It is in this manner, use case upon use case, that you will be able to track your digital maturity over time. I think the best lesson from projects are what gaps you see that you could push toward for your next project. This way you will start to introduce a competitive edge as you push beyond the standard implementations.

Now we have established the general workflow, the next chapter will build upon it for a more complicated example, using more data sources.

## CHAPTER 8

Use Case: Audience Segmentation

This use case follows the same format as the Predictive Audiences GA4 features introduced in Chapter 7 but extends to a more complicated example.

In this scenario, suppose our book publishing company has had some great success using Predictive Audiences and has seen revenue increases as a result, so we now have been invited to put together a budget proposal to explore if more intelligent targeting would benefit the business. To help with this, we have access to an internal CRM database with a large history of purchases along with the customer's profession given at time of signup. We want to see if we can use this extra data as to improve relevance in the book advertisements they see. The expectation is that if, say, a customer is a doctor they would appreciate ads about medical books and convert more often. Repeating this personalization for many professions, we anticipate an overall increase in conversions and therefore revenue for the publisher.

For this use case, we will create segmentation of our users using the CRM data and then make it available to our GA4 event data as it streams in. We'll store the data within GCP and join the data using BigQuery; we'll use Firestore and GTM SS to help merge the data with GA4 events if relevant, and then activate by creating GA4 Audiences for use within the Google Marketing Suite. However, we first, as always, start with the business case to justify the project.

Creating the Business Case

The overall aim is to segment your customers so you can create better experiences for them. We would like to achieve more efficiency in our Google Ads costs, which will be our data activation channel. A future plan would be to use the same segment for other channels such as Google Optimize. Our business case is to reduce costs and increase conversion to get higher sales as a result of tailoring our messaging more tightly to the customer.

Assessing Value

Assessing the value is similar to what we did in the previous chapter, "Assessing Value" on page 236—as a reminder we ended that chapter with the revenue and costs shown in Example 8-1.

Example 8-1. Accountancy numbers when the predictive audiences use case is enabled (actual)

```
Google Ads monthly budget: $10,000
Cost per click: $0.50
Clicks per month: 20,000
Conversion rate: 10%
Average order value: $500
Orders: 180 (top 9% who buy anyway) + (1800 * 5% uplift) = 2070

Monthly Revenue: 2070 orders * $500 = $1,035,000
Monthly ROI: $1,035,000 / $10.000 = 103.5

Actual uplift: $35,000 a month
```

This time, however, we're going to look at having extra dimensions associated with each customer, such as customer lifetime value and profession. We will then use that data to create more segments, subsetting the Predictive Audiences we made in Chapter 7. We'll then be able to generate Audiences to create doctors, teachers, builders, writers, etc., which can be used for more fine-grained ads aimed at those customers.

Given our experience with predictive audiences in Chapter 7, we should be better informed about our potential results. Before, we estimated a 10% uplift but actually got only 4%. Let's say our uplift estimate from when we tried out predictive audiences is less ambitious, namely, an extra 5% uplift for all segments we target, as shown in Example 8-2. This translates to an estimated incremental increase in revenue of $51,500 a month.

Example 8-2. Estimated accountancy numbers for segmentation use case

```
Google Ads monthly budget: $10,000
Cost per click: $0.50
Clicks per month: 20,000
Conversion rate: 10%
Average order value: $500
Orders: 2070 * 5% uplift from segmentation = 2173

Monthly Revenue: 2173 orders * $500 = $1,086,500
```

```
Monthly ROI: $1,035,000 / $10.000 = 108.65

Estimated uplift: (estimated:$1,086,500 - now: $1,035,000) = $51,500 a month
```

From this we can now estimate the incremental value of our project. We are going to spend resources on creating this uplift—is it worth it? If we hit our estimates, we should see an extra $51,500 monthly in revenue on top of the predictive audiences, but we will have more cloud costs (*https://oreil.ly/YinMM*). The actual cost will depend on how often you train your model, the services you use, how much data you have, and any ongoing babysitting tasks people need to do.

There will also be implementation costs. It's common to look at an upfront investment of cost versus time to earn that investment back as a way to judge its effectiveness. If, say, implementation costs $200,000 to implement, then an extra $51,500 per month means it will take only four months to earn it back, which is usually very acceptable. You are, however, always taking on some technical debt when you implement your own solution, so factor in days or hours per month to maintain it plus the running cost itself. In summary, you may combine all of this into a hard cost limit that the solution needs to pay back within three months/six months/one years worth of incremental benefit to be worthwhile. An example could be that you expect a profit within one year, so it should not cost more than $51,500 × 12 months = $618,000 to enable.

If it's not worthwhile, then this is valuable information. Perhaps you can then work out the percentage uplift you will need for a project to be worthwhile (say, 20%) or a limit on implementation costs to properly assess future use cases. With a rough figure in mind on what we can spend, let's turn to what cloud resources we'll use.

Estimating Resources

Knowing which resources you will need takes experience with the different aspects of the technologies involved, which is one of the aims of this book. Nevertheless, weighing up capabilities verses cost is a nontrivial task. I suggest you first design something that simply works within the scope you have created and try not to over-optimize too soon. Look also to use as few moving parts as possible, meaning keeping the number of technologies low. Once you have a working solution, you can look at optimizing its costs and features. I'm going to jump right in with a suggested solution using the resources outlined earlier in the book:

GA4

You'll need to configure GA4 to provide a userId for linking the web activity to the BigQuery CRM imports. You usually do this via a login screen or form that allows you (after consent) to link the two data sources. For this example, we will assume that the CRM and GA4 are capturing a common userId generated from the website content management system (CMS) when a user logs in. Our example data has this in the format "CRM12345." We also assume that we have a good percentage of users logging in to make a purchase when a login is mandatory.

GA4 userId to Link to Your CRM

If a userId via a login area isn't available for your website, fostering such a dataset may require a whole new website strategy. In modern times, with cookie restrictions and user privacy paramount, the value of reliable user profile data is always increasing. Gaining that data from your users requires building up a trusted brand with incentives to encourage them to give you their data.

BigQuery

You'll need to link BigQuery (see "BigQuery" on page 109) to the GA4 property. This will contain the GA4 BigQuery export plus the extra CRM database imported from your other systems. One of the key requirements for this use case is a way to link the user data in your CRM with the web analytics data in GA4. See "Linking Datasets" on page 175 for things to consider when doing this. Within BigQuery, we can also extend the use case to use BigQuery ML to add some machine learning metrics to your data, such as predicted lifetime value or likelihood of churn.

CRM Imports into BigQuery

We will also need someone who is familiar with the internal CRM system to create the exports and schedule them for import into BigQuery. To make the role easier, you can ensure they are responsible for only the exports into GCS (see "GCS" on page 122), whereas a cloud engineer role will take over importing that data into BigQuery. You need good communication between these roles to make sure the exports are suitable.

Cloud Composer

Once we have the SQL for creating the BigQuery table of joined GA4 data and CRM data, we'll use Cloud Composer (see "Cloud Composer" on page 131) to schedule the daily updates. This could be swapped for a scheduled query within BigQuery, but we'll also be using Cloud Composer in the next step to send the data to Firestore, so it's more convenient to keep them both within the same systems.

Firestore

To use BigQuery in real time as a user browses the website, we'll move the BigQuery data into Firestore so it will be available immediately. You can set this to

---

run daily, weekly or hourly depending on how often you anticipate the data will change. Daily will be fine for our example.

GTM SS

With the Firestore connector, you will use GTM SS to enrich the GA4 data streams with our model data as a user browses the website. This will add the CRM data as it currently stands within Firestore, which will then be sent to the GA4 UI ready for exporting via GA4 Audiences, as we did in Chapter 7.

Overall, the cloud running costs are estimated to be the following:

- BigQuery daily queries: $100 a month (depending on data volume)
- Firestore read/write updates: $100 a month (depending on how the number of calls and updates)
- GTM SS: $120 a month (assuming standard App Engine set up)
- Cloud Composer: $350 a month (may be reused for other projects)

Altogether, this looks to be around $670 a month for the data volume in our example. This will differ for your own business, but something in the region of budgeting $500 to $1,000 a month should be in the ballpark. Referring back to our estimated uplift of $51,500 calculated in Example 8-2, we are comfortable that we should have an overall positive revenue impact for the project, even if we add on set-up fees and an employee role to look after the solution.

We have "known unknowns" regarding how much exact data volume will be passed between systems, so I try to round upward as much as possible at this stage. It's rare that these cloud costs are prohibitive, and we need to turn cloud services off again once we measure the results. We should be prepared for that eventuality, and these figures will help in those decisions. Note here that if you have a requirement to move data outside of GCP (say, to AWS), these costs can become significant. Also beware of Cloud Logging costs that can contribute hundreds of dollars to the monthly figure— you may want to turn these off after the development phase of the solution.

Now that we have a rough idea of the costs and value of the project, let's move on to (my favorite bit) putting all the pieces together.

Data Architecture

This section covers how the different GCP services will interact with one another and will both help you clear up what exactly needs to be done and help you report to other stakeholders what it is you're working on. These data architecture diagrams will serve as a first point of call for documentation of the service, which is essential if you're planning to pass on the system maintenance to someone else in the future. The data architecture diagram in Figure 8-1 represents my own third or fourth iteration of the system as I was writing this chapter, as I considered the use case and technologies I wanted to write about. I imagine you will have the same experience as you brainstorm possible solutions for your own business.

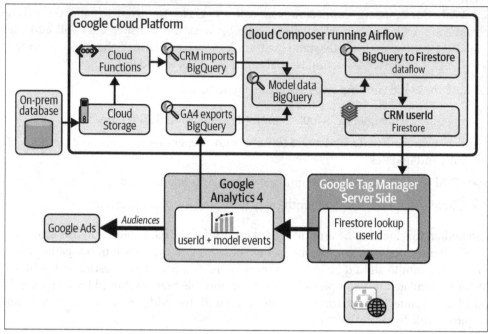

Figure 8-1. Data architecture for the user segmentation use case

With the data architecture in place, our job now is to essentially create the configuration and set it up to enable the nodes and edges within the diagram. The first priority is how data will flow in via GA4 and the CRM system.

Data Ingestion

We move on to configure how the GA4 data and CRM data will end up in BigQuery, since this will be where we make the join between the two systems. Now that we're not using the GA4 interface, one of the first steps will be to configure GA4's data collection and ensure that the correct data appears in GA4's BigQuery export.

GA4 Data Capture Configuration

As our use case will be working with individual user journeys, we're going to need a way to link our CRM data and the website user's activity. This means linking via a userId, and we will need to have the granular level of data that the GA4 BigQuery export provides.

We're now working with pseudonymous data and identifiable user data, so we need to ensure we're in privacy compliance. To allow for such a link between systems and to use a person's identity and data, I recommend getting explicit user consent that allows for using their data to target them with more relevant content.

For our scenario, we assume we have a user login area that a good proportion of your users are using for enhanced website features. This will facilitate the use of the `user_id` variable within your GA4 dataset. You may also have some additional `user_properties`, such as consent status, which we'll also add to via a `profession` property that will hold the result of this use case when it's imported back into GA4. We talked about how to work with GA4 by adding user properties in Chapter 3 ("User Properties" on page 61).

To see the extra user profession in the GA4 reports, we'll need to configure a Custom Definition at User level scope, which will be populated within GTM on the way into GA4, as seen in Figure 8-2.

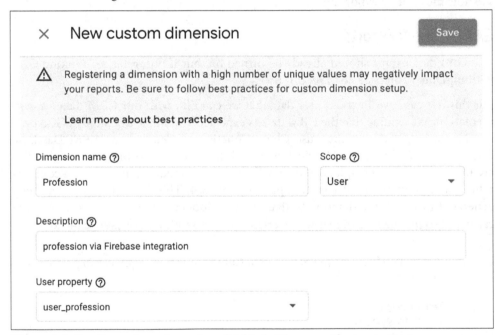

Figure 8-2. Configuration of a custom field to hold user profession in GA4

To populate the standard `user_id`, the data will need to be sent from the website via either `gtag.js` or Tag Manager dataLayer pushes, as shown in Example 8-3. We show the latter as it's the recommended way to implement GA4 tracking, taken from code examples from the Google documentation for sending user IDs (*https://oreil.ly/Lg249*).

Example 8-3. A dataLayer push to send data to GA4 with CRM IDs, typically populated when a user logs into a website via a web form

```
dataLayer.push({
  'user_id': 'USER_ID',
  'crm_id': 'USER_ID'
});
```

We will fill in the custom `user_profession` field later via GTM SS.

With the `userId` and the custom fields of the added data you want to import in place, we should have enough GA4 configuration to collect our enhanced data. Later in the project data activation phase, we'll come back to GA4 configuration to create the Audiences.

We now move on to exporting data into BigQuery.

GA4 BigQuery Exports

The GA4 data export should already be turned on, but if not, refer to "Linking GA4 with BigQuery" on page 76 to set this up.

For this use case, we'll extract user data that we can link with our CRM data for segmentation. We assume that the CRM data is exported and made available in GCS on a regular basis, so we can then use Cloud Function as detailed in "CRM Database Imports via GCS" on page 94, which will load that data within BigQuery as the files are made available. The uploaded data will be similar to that example, with a configuration file to trigger Cloud Function per Example 8-4. This fake CRM data has been generated to match the demo GA4 data for the Google Merchandise Store in some cases and so includes a `cid` that will overlap with GA4's BigQuery `user_pseudo_id`.

Example 8-4. A YAML config file for use with Cloud Storage to BigQuery import Cloud Function specified in Chapter 3

```
project: learning-ga4
datasetid: crm_imports_us
schema:
 fake_crm_transactions:
  fields:
    - name: name
     type: STRING
    - name: job
     type: STRING
    - name: created
```

```
  type: STRING
- name: transactions
  type: STRING
- name: revenue
  type: STRING
- name: permission
  type: STRING
- name: crm_id
  type: STRING
- name: cid
  type: STRING
```

I then generated the CSV files as would a real-life CRM export and placed those within the Cloud Storage bucket that the Cloud Function trigger was configured to import from. It imported the data and made it available within the BigQuery table `learning-ga4:crm_imports_us.fake_crm_transactions`, as shown in Figure 8-3.

Row	name	job	created	transactions	revenue	permission	crm_id	cid
1	Jannette Walsh DVM	Sub	2010-11-23 00:14:58	73	9425.59	TRUE	CRM000040	54318914.1826922613
2	Esther Schmitt	Sub	2007-07-30 20:57:12	167	10036.49	TRUE	CRM000372	6473678.0978223839
3	Stevan Kertzmann	Sub	2010-10-14 17:06:29	65	9286.62	TRUE	CRM001727	45682856.7032608942
4	Mr. Hoy Rosenbaum	Sub	2009-08-17 14:17:20	71	7995.74	TRUE	CRM001920	27591007.2215776243
5	Lainey Schneider-Bailey	Sub	2018-04-23 13:28:43	103	2851.85	TRUE	CRM002273	3681170.7474480108
6	Dr. Jovany Hilll DDS	Sub	2007-04-24 08:18:01	414	5004.34	TRUE	CRM003647	59175216.1880512930
7	Adele Larkin-Murazik	Sub	2011-04-06 06:56:14	384	17003.58	TRUE	CRM004028	70205933.0713603324
8	Sal Blanda	Sub	2011-03-07 00:37:33	70	8854.83	TRUE	CRM004645	49495289.2240771945

Figure 8-3. Fake CRM data within BigQuery generated to overlap the Google Merchandise Store cookie IDs

We assume that the `cid` value for this dataset came from the website when those users logged in. The `cid` value for the GA4 cookie is read in and added to the HTML form as a hidden form entry, which is read into the CRM system.

It is likely that users will become associated with many `cid` values if they clear cookies or use different browsers. The nature of web data is that it is normally pretty messy, sadly. It will represent some errors in your dataset, but how much of an issue this will be depends on how often your users log in and what your privacy controls are.

By the end of this phase in a real-life case, you should have two datasets within Big-Query. To help you walk through this use case yourself even if you don't have the same data, I've used the public BigQuery example dataset for GA4 to create public datasets for this book:

- The GA4 data export dataset named something like `analytics_123456` in your own case is simulated via the public GA4 dataset for Google Merchandise Store at `bigquery-public-data.ga4_obfuscated_sample_ecommerce.events_*`.

- For your CRM data import, in, say, a dataset called `crm_data`, I have created an example dataset with some cookie IDs that overlap the sample data. All names and professions are randomly generated (in R via the `charlatan` (*https://oreil.ly/Ojih9*) package) and available for you to query via `learning-ga4.crm_imports_us.fake_crm_transactions`.

Assuming your data is ready to use (or you're ready to use the sample datasets), let's move to the next phase of consideration, how the data will be stored.

Data Storage: Transformations of Your Datasets

By now your raw data of GA4 and your CRM datasets should be imported daily into BigQuery, but this data is rarely in a useful form just yet. It's beneficial to create tidy, aggregated forms of the data, and these will serve as the "source of truth" for further derived data flows, as described in "BigQuery" on page 109.

For this use case, we want to create a dataset suitable for export to Firestore. This will merge the CRM and GA4 data on the `userId` so that each `userId` has a row of both online and offline fields in the table. There will likely be some back-and-forth as you include and discard fields within the dataset, so a work environment where you can iterate quickly is useful, which the WebUI of BigQuery provides.

I would usually start by creating an aggregated table with just the GA4 ecommerce data. The query Example 8-5 shows an example you could run each day to populate such a table. Using this small example, it doesn't really matter if this is called directly within the join later, but for real-life use cases with lots of data, it will be better to make this query populate an intermediate table that will probably save you in costs.

Example 8-5. SQL to get transaction data from the demo GA4 dataset in BigQuery

```
SELECT
 event_date,
 user_pseudo_id AS cid,
 traffic_source.medium,
 ecommerce.transaction_id,
 SUM(ecommerce.total_item_quantity) AS quantity,
 SUM(ecommerce.purchase_revenue_in_usd) AS web_revenue
FROM
 `bigquery-public-data.ga4_obfuscated_sample_ecommerce.events_*`
WHERE
 _table_suffix BETWEEN '20201101' AND '20210131'
GROUP BY
 event_date,
 user_pseudo_id,
 medium,
 transaction_id
HAVING web_revenue > 0
```

The query should be available for you to run in your own BigQuery console since it is using public data. When it's executed, you should see results similar to those in Figure 8-4.

Figure 8-4. The results of a transaction query upon the public GA4 data

With our two datasets ready for joining, we can move on to creating the join and exporting the data ready for Firestore.

Data Modeling

We are now onto the third phase of our data project as covered in Chapter 5. For this use case, our modeling will be a simple join between our two datasets, which although simple will help demonstrate that this kind of linking of datasets is often powerful. We could also easily extend the use case for other goals, since within Big-Query we could use BigQuery ML to add machine learning metrics on top of the data, such as predicted lifetime value.

For our purposes, we need to add the profession dimension sourced from the CRM system and link it to the latest GA4 cid value. An example of how this join could work is shown in Example 8-6.

Example 8-6. SQL to merge transaction data from the demo GA4 dataset in BigQuery with some fake CRM data created for this book

```
SELECT crm_id, user_pseudo_id as web_cid, name, job
FROM
 `learning-ga4.crm_imports_us.fake_crm_transactions`
 AS A
INNER JOIN (
 SELECT
  user_pseudo_id
 FROM
  `bigquery-public-data.ga4_obfuscated_sample_ecommerce.events_*`
 WHERE
  _table_suffix BETWEEN '20201101'
  AND '20210131'
 GROUP BY
  user_pseudo_id) AS B
ON
 A.cid = B.user_pseudo_id
ORDER BY name
```

Running the query should give you a result similar to Figure 8-5.

This data is not ready yet, however, because we need to import it into GA4 somehow, which we will do with Firestore.

```
 1  SELECT
 2    crm_id,
 3    user_pseudo_id AS web_cid,
 4    name,
 5    job
 6  FROM
 7    `learning-ga4.crm_imports_us.fake_crm_transactions` AS A
 8  INNER JOIN (
 9    SELECT
10      user_pseudo_id
11    FROM
12      `bigquery-public-data.ga4_obfuscated_sample_ecommerce.events_*`
13    WHERE
14      _table_suffix BETWEEN '20201101'
15      AND '20210131'
16    GROUP BY
17      user_pseudo_id) AS B
18  ON
19    A.cid = B.user_pseudo_id
20  ORDER BY
21    name
```

Processing location: US

▶ Run ▾   ⬇ Save query   ⬚ Save view   🕐 Schedule query ▾   ⚙ More ▾

Query results

⬇ SAVE RESULTS   📊 EXPLORE DATA ▾

Query complete (7.9 sec elapsed, 94 MB processed)

Job information **Results** JSON Execution details

Row	crm_id	web_cid	name	job
1	CRM120083	48862427.9017261510	Aaden Aufderhar	Scientist, physiological
2	CRM064568	87857566.7483865271	Aaden Brakus	Accountant, chartered
3	CRM006163	31877872.8352061999	Aaden Fay	Ceramics designer
4	CRM111232	30235785.2184113721	Aaden Haley	Community development worker
5	CRM087542	8578822.9947669395	Aaden Hayes	Surveyor, land/geomatics
6	CRM035641	62943588.5643621894	Aaden Heaney-Wisoky	Local government officer
7	CRM122462	9210276.8861027889	Aaden Hoeger	Estate agent
8	CRM056689	13438757.8854190306	Aaden Johnston	Astronomer
9	CRM069852	65136379.8546335808	Aaden Kassulke DDS	Theatre director

Figure 8-5. An example showing the result of joining the demo datasets from GA4 and CRM

Data Activation

We'll now look at the steps to activate our joined data, currently sitting in BigQuery.

To get the CRM data where we need it, enriching the GA4 data stream as it is sent to GA4, we need to intercept the GA4 call as the user browses the website. GTM SS provides this functionality via its Firestore variable. We first need to get the data into Firestore so GTM can read it.

The BigQuery table has been exported with the crm_id (see Figure 8-5), which we will use as a key for Firestore.

The query outlined in Example 8-6 will be saved to local file ./join-ga4-crm.sql that Cloud Composer will use as an input to its starting task. That table will then be passed to Firestore using the BigQuery to Firestore Cloud Composer DAG, as detailed in Example 8-7.

Example 8-7. A Cloud Composer DAG to create a BigQuery table and send it to Firestore via Dataflow. In this case, the BigQuery SQL to create the BigQuery table with one column containing a userId is assumed to be in a file named ./join-ga4-crm.sql.

```python
import datetime
from airflow import DAG
from airflow.utils.dates import days_ago
from airflow.contrib.operators.bigquery_operator import BigQueryOperator
from airflow.contrib.operators.gcp_container_operator import GKEPodOperator

default_args = {
    'start_date': days_ago(1),
    'email_on_failure': False,
    'email_on_retry': False,
    'email': 'my@email.com',
    # If a task fails, retry it once after waiting at least 5 minutes
    'retries': 0,
    'execution_timeout': datetime.timedelta(minutes=240),
    'retry_delay': datetime.timedelta(minutes=1),
    'project_id': 'your-project'
}

PROJECTID='learning-ga4'
DATASETID='api_tests'
SOURCE_TABLEID='your-crm-data'
DESTINATION_TABLEID='your-firestore-data'
TEMP_BUCKET='gs://my-bucket/bq_to_ds/'

dag = DAG('bq-to-ds-data-name'),
    default_args=default_args,
    schedule_interval='30 07 * * *')
```

```
# in production sql should filter to date partition too e.g. {{ ds_nodash }}
create_segment_table = BigQueryOperator(
  task_id='create_segment_table',
  use_legacy_sql=False,
  write_disposition="WRITE_TRUNCATE",
  create_disposition='CREATE_IF_NEEDED',
  allow_large_results=True,
  destination_dataset_table='{}.{}.{}'.format(PROJECTID,
                                              DATASETID, DESTINATION_TABLEID),
  sql='./join-ga4-crm.sql',
  params={
    'project_id': PROJECTID,
    'dataset_id': DATASETID,
    'table_id': SOURCE_TABLEID
  },
  dag=dag
)

submit_bq_to_ds_job = GKEPodOperator(
  task_id='submit_bq_to_ds_job',
  name='bq-to-ds',
  image='gcr.io/your-project/data-activation',
  arguments=['--project=%s' % PROJECTID,
          '--inputBigQueryDataset=%s' % DATASETID,
          '--inputBigQueryTable=%s' % DESTINATION_TABLEID,
          '--keyColumn=%s' % 'userId', # must be in BigQuery ids (case sensitive)
          '--outputDatastoreNamespace=%s' % DESTINATION_TABLEID,
          '--outputDatastoreKind=DataActivation',
          '--tempLocation=%s' % TEMP_BUCKET,
          '--gcpTempLocation=%s' % TEMP_BUCKET,
          '--runner=DataflowRunner',
          '--numWorkers=1'],
  dag=dag
)

create_segment_table >> submit_bq_to_ds_job
```

In production, this is scheduled every day to import data, either overwriting updating values or creating new fields as necessary. This data is then available for the real-time stream as the users browse the website, sending GA4 hits. Once the BigQuery CRM data is imported into Firestore, you should see your data with the CRM keys, as shown in Figure 8-6.

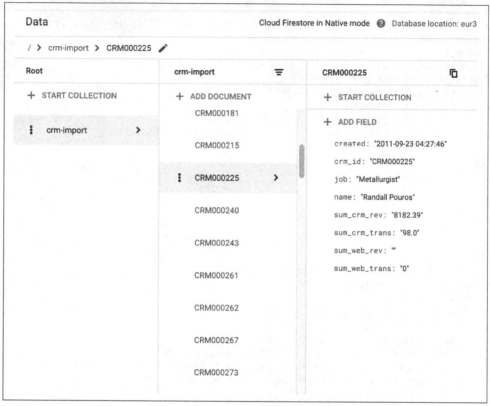

Figure 8-6. Your CRM data imported into Firestore from BigQuery

Now we have a daily updating BigQuery join with CRM data that is exported to Firestore. The last remaining step is to merge that data with GA4 web streams when a user with a recognized cid visits the website, which we'll accomplish with GTM SS.

Setting Up GA4 Imports Via GTM SS

We now look at how to set up your GTM SS instance to fetch data from Firestore and add it to your GA4 stream. We won't cover how to set up GTM SS as detailed in "GTM Server Side" on page 80, but it's assumed to be a standard App Engine instance that costs around $120 a month in cloud costs. Note that we use GA4 here, but you could extend this example to also send data to other digital marketing services from the same events.

One option for this stage could have been to import into GA4 via its Data Import (*https://oreil.ly/PGpmI*) service—and indeed that's what I first intended—but at the time of writing this is available only as a manual process. It may be an easier option once it's available via an API or BigQuery connector.

However, because Data Import is not available, and I would also like to consider more real-time applications, we'll instead go the route of BigQuery to Firestore to GTM SS. This has more potential applications due to its real-time nature as well.

This workflow is facilitated by the new Firestore variable, which makes this flow much easier than it used to be: we need only point our Firestore variable within GTM SS and then populate the GA4 server-side tag with that data.

Referring back to our GA4 configuration in "Data Ingestion" on page 250, we need to add a User Property called `user_profession` to our GA4 tag, using the `user_id` field as the document name.

When using GTM SS, your web container will be set up to send GA4 events to your own URL at, say, *https://gtm.example.com*. These GA4 events will contain all the fields as configured in your web GA4 tag, including `user_id`. We'll first need to extract that in a GTM variable so we can use it as a document name to fetch from Firestore—that configuration is shown in Figure 8-7.

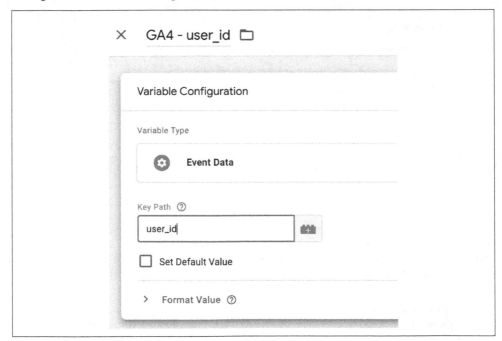

Figure 8-7. Setting up a custom event to extract the `user_id` we will use as the document name to fetch data from Firestore

Next we set up the Firestore Lookup variable to use that `user_id` variable (accessed via `{{GA4 - user_id}}`) as our Document Path for Firestore. Document Paths are in the format *{firestore-collection-name}/{firestore-document}*; in this case, the *firestore-collection-name* will be what you called it when setting up Firestore, and *firestore-document* will be the CRM ID we can see in Figure 8-6 that we uploaded. We can also see that within that document the key will be "job" because that is the name of the field within the Firestore document. See Figure 8-8 for an example of this configuration.

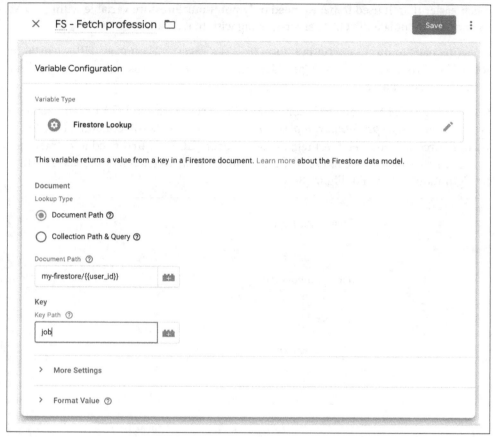

Figure 8-8. Configuring a GTM SS Firestore Lookup variable to call our Firestore collection containing the CRM data, using user_id *as its document reference*

If we want to reach other data within Firestore, we would repeat the process to reach other values such as "name" or "crm_web_rev". You can also fetch nested records.

Finally, we create a GA4 tag that will populate the GA4 event to send to our GA4 account. See Figure 8-9. This should be configured to fire on the GA4 events when you send in the user_id, such as a login Recommend Event.

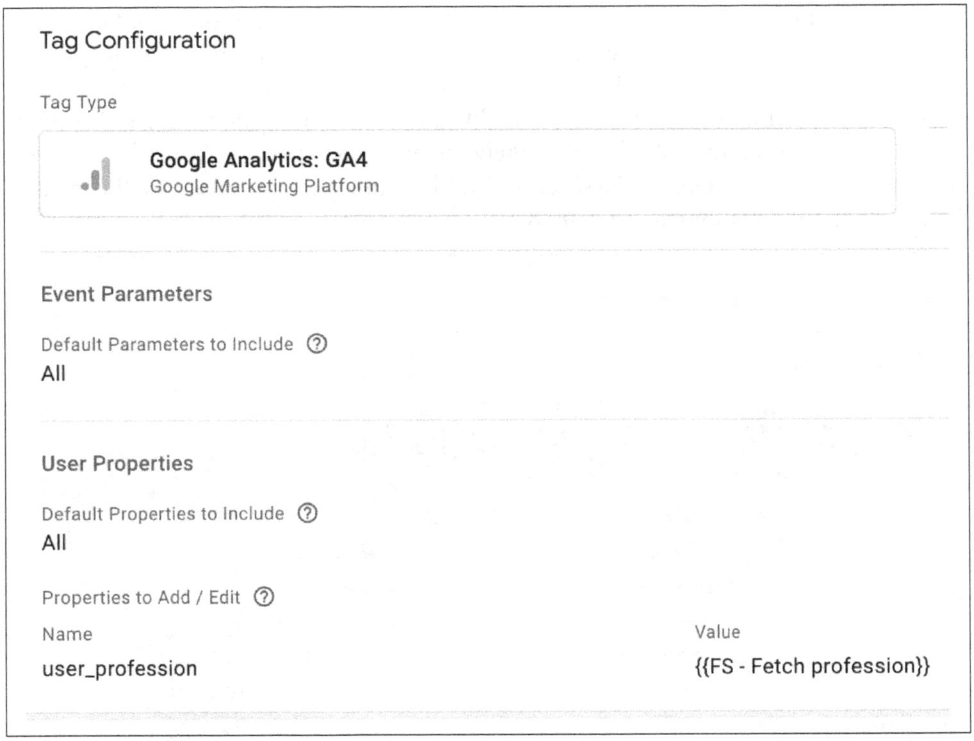

Figure 8-9. Configuring a GA4 GTM SS event tag with user properties adding the Firestore value

The `user_profession` parameter should match the Custom Field configured earlier via Figure 8-2, as it's this event parameter that will be referenced to populate the GA4 custom dimensions and enable it to be available for Audiences.

After everything is published and tested, you should see your GA4 custom dimension start to populate with data when a recognized user logs in with the same CRM ID as has been uploaded to BigQuery, and so on to Firestore. You will probably need a couple of weeks of testing to ensure that your user IDs, schedules, and matching is adequate and to quality-check the configurations. All things going well, after a couple of weeks you should then start to have the data available to begin working with those customers who have shared their professions with you, and we can activate them in a similar manner as the previous use case within our Predictive Audiences.

Exporting Audiences from GA4

Once you can see user professions within your GA4 data, you will be able to start using them in all reports, such as Explorations and the real-time reports. They will also be available for Audiences.

For this example, we wish to build upon the Predictive Audiences as created in Chapter 7, only this time we'll create more sub-Audiences that will target certain professions. Let's say we know that doctors, teachers, and builders are key professions for us, so we start with those, although in reality this list will be much larger. We then aim to create Audiences of "Doctors likely to purchase in next 7 days," "Teachers likely to purchase in next 7 days," and "Builders likely to purchase in the next 7 days"—an example is shown in Figure 8-10.

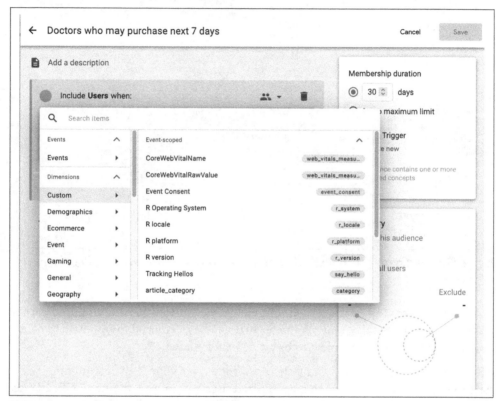

Figure 8-10. Adding a new custom dimension to our Audience definitions that will combine with the existing Predictive Audience

These audiences will combine with Predictive Audiences, as we shown again in Figure 8-11.

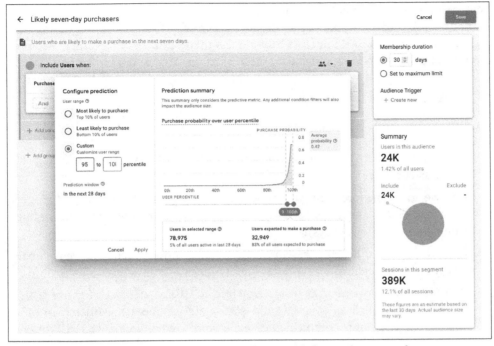

Figure 8-11. Configuration of an Audience showing likely purchasers in the next seven days

Once the sub-Audiences are configured, they will accumulate users over time, and these Audiences can be exported to Google Ads and other services within the Google Marketing Suite as previously detailed. It would then be a matter of handing over these Audiences to your Google Ads marketing team, who can create targeted ad copy or suppress or tailor campaigns for them.

Testing Performance

The general principles mentioned in "Data Activation: Testing Performance" on page 242 apply here also, namely, A/B testing your Audiences so you can get an incremental increase over your baseline will be the most powerful way to demonstrate value for your data project in general.

I should also mention that although focusing on a use case's business case is the best way to get the project delivered, there are many side benefits you will start to appreciate once you've gone through implementation a few times. The tech stack for this use case alone gives you a solid launching platform for many others, ones that will be easier to quickly bring online since the sunk costs of the technology needed for this use case (e.g., Cloud Composer, GTM SS, the Firestore database) are already paid for.

With more Audiences, you'll find a lot more actions and insights during your performance testing, and I can imagine a whole year's worth of follow-up use cases could result. For example, why do doctors respond less than builders to targeting? Are predictions of conversion more reliable for teachers as opposed to doctors? And so on.

Summary

I would investigate the natural next steps:

- Can we add other dimensions from our CRM data that we think will help the business/customer?

- Can we add some forecasting or modeling at the BigQuery stage using BigQuery ML to make more intelligent segments (such as predicated lifetime value, not just value in the next seven days)?

- Can we activate the Audiences within Google Optimize to change website content to help conversion?

- Can we also export metrics from GA4 into the CRM database for use within other channels such as email? Could customers get the same offers across all channels for a more omnichannel experience (email, Google Ads, on-site banners)?

Most of these extensions involve updating the configurations for the existing tech stack we deployed in the use case.

I hope this chapter has helped bring together a lot of the concepts the rest of the book has been leading up to. We've covered the strategizing and justifying the use case, ingesting data from GA4 and a CRM system, transforming the data and modeling in BigQuery, and then making it available for activation via Firestore and GTM SS onward to Google Marketing Suite.

The project outlined is similar to several I have carried out in my career, and I've seen big benefits to their implementation both in terms of direct monetary value and the digital transformation and energy it adds to a company. As mentioned, it will be difficult to copy over exactly because every business is different, but hopefully the generic components can spark your own creativity and help you apply it to your situation.

This use case has worked on an activation channel of Google Ads, but the next chapter will look at another activation channel in a real-time environment: creating a real-time dashboard from GA4 data.

## CHAPTER 9

Use Case: Real-Time Forecasting

This use case moves into real-time data flows, so we'll look at utilizing the Real-Time GA4 API. We'll also include some real-time actions on our website using the Audiences we created in Chapter 8 via Google Optimize.

In this scenario, let's assume that our book publishing company has now been using the predictive, segmented audiences we created in previous chapters for six months and has found them useful in increasing relevance, so conversions are up for certain sectors. The social media marketing team have learned about this in an internal presentation and have asked if they can use the same segments to improve their everyday activity posting promotional content on the brand social media accounts. We propose that if the social media team could see the real-time effects of content and which Audience they resonate with, they could quickly respond with more follow-up content of the same nature. As they often post topical news content, waiting for analysis the day after may be too late to take action on the data.

It would be most helpful for the content team to know what content is popular before it is—a forecast of engagement trends will help you estimate what will be "hot or not." We would also like to set up a process where content a user sees in social media campaigns will also be present on the website, personalized to their profession. The team needs an easy way to update this content, which we opt to do via a Google Optimize banner (see "Google Optimize" on page 201) that will trigger at the top of the page.

Creating the Business Case

We want to increase traffic to the website by increasing the relevance of the social media content feed, and increase conversions by also showing a banner that will shortcut to our promotion pages connected with each social media campaign. Forecasts of the current trends for content will help us know what content to prioritize for this banner.

Social media activity is considered "upper funnel" in that most users won't convert immediately, but it helps keep the publisher brand in mind when customers are considering purchases later. KPIs for the social media team are then largely based on impressions and engagement with their posting, and they wish to increase these metrics via monitoring the trends on the dashboard.

Once on the website, the banner should help directly with increasing conversion rate as it will shortcut navigation paths to relevant landing pages. We expect more conversions as a result. For example, if a particular medical procedure to help long-term sufferers of COVID-19 is hitting headlines across the world, then content related to virology books could be more relevant and attract unseasonal traffic. Seeing this trend, you can prioritize content promoting virology books. Once a medical practitioner hits the website (as identified by our Audiences), then that particular book could be highlighted in the banner because it's likely to be they're looking for.

Resources Needed

As stressed in "Making Dashboards Work" on page 203, we need to ensure that sufficient action will be enabled when using a dashboard. This inserts the role of a human being within your data flow, whose job it will be to react to the data and make decisions based on that information. There is no point to having a real-time dashboard unless you can make real-time decisions.

The key decisions are scoped to be as follows:

- Which social media content should be prioritized for publication?
- What content should be presented to each Audience in a banner once they hit the website?

This role needs to be able to operate Google Optimize because that will be the vehicle for creating the banners, which having some HTML frontend skills would help with. We choose Google Optimize banners because of the ease of use and speed to change content once they have been set up.

While the most important piece of this is having the right person in this role, we'll also support them by using these technologies:

GA4

The setup includes previously created Audiences for each profession like we did in Chapter 8. We'll be fetching event names and Audience name from the real-time API.

R Shiny

This will take the dashboard role, since R can also handle the forecast modeling, the GA4 real-time API calls, and the interactive presentation layer. It's hosted with GCP. This role could be assigned to any other dashboard system that has forecast capabilities.

Google Optimize

This is the platform used to create the HTML banner on the website. It's linked to GA4 so it can tailor each banner to each of the Audiences imported, such as the profession of the user if that is recognized.

With all these components identified, we'll put them together in the next section.

Data Architecture

The data architecture for this solution includes an unusual component: a human being. This highlights the key importance of the decision making that only a human can provide and that we can't (yet) replicate with an automated system. The major decision branch points are highlighted in Figure 9-1, which will determine which content is promoted on the publisher's social media and which content can be added to Google Optimize.

 In the future, if we discover this role includes repeatable tasks, perhaps we can automate more in a follow-up project. A potential route is using some of the machine learning APIs we spoke about in "Machine Learning APIs" on page 181.

With our plan in place, we now look at how to connect the dots in our architecture.

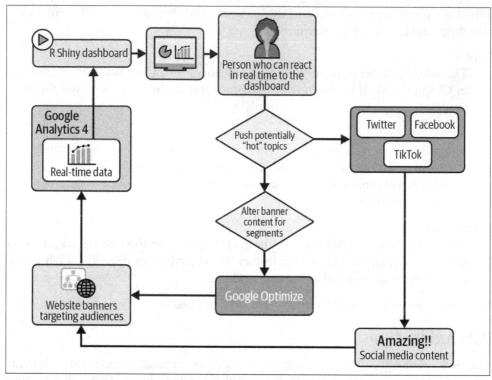

Figure 9-1. Real-time data is taken from GA4 and a forecast is created to help prioritize content for social media and on-site banners via Google Optimize

Data Ingestion

The first thing to consider is how data will enter the system, as we covered in Chapter 3. For this use case, this means making sure the GA4 real-time stream will hold all the information we need for the dashboard.

GA4 Configuration

The data we want to pull in real time needs to be listed within GA4's Real-Time API dimensions and metrics (*https://oreil.ly/w2rw6*). This is a limited subset of the dimensions you can get via the normal API. By default, you can get the last 30 minutes of event data from your website, or you can get 60 minutes with a paid GA360 license.

For our purposes, we're looking to see how many users are on the website from the Audiences we have created and which section of the website they're on. This will allow us to see which content is resonating best within the Audience segments we have created. Referring to the Real-Time API, we see that the fields `audienceName` and `unifiedScreenName` can identify the data we need. To get a forecast, we also need a time-series trend, so `minutesAgo` will be used to order the user activity and extrapolate forecast trends.

However, there are further restrictions on the real-time API that will restrict which dimensions and metrics can be queried together. This includes trying to query `audienceName` and `unifiedScreenName` (e.g., a page URL) together. This is related to trying to tie a user-based metric to an event-based metric. We can always query `event_name` and `event_count`, which is the fundamental data schema of GA4, so to make the data we want as easy to extract as possible, we may wish to consider having events that will record the user interactions we're looking for. You can also query user-scoped Custom Dimensions (but not event-scoped) as described in "User Properties" on page 61.

Note also that while you cannot query `audienceName` and `unifiedScreenName` together to see, say, doctors reading medical content, you can define an Audience with page-name characteristics.

An example for the setup for this use case is shown in Figure 9-2. We've created Audiences that will populate when a user profession matches content we want to monitor, namely doctors reading medical content.

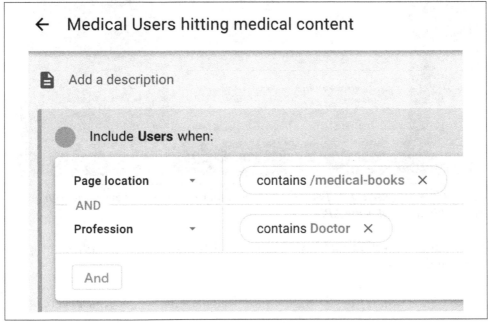

Figure 9-2. Creating an audience to query in the Real-Time API matching medical books to users who are doctors

When users hit this audience, they will also trigger an event we can query, so this could be something to query in the Real-Time API, too. We could set up an event called `doctors-seeing-medical-content`, for example, as in Figure 9-3.

We'll go with this last option, to keep it familiar to GA4 users: the real-time dashboard will reflect the Audiences as configured in GA4. This provides the easiest way for users to add or remove data from the real-time feed.

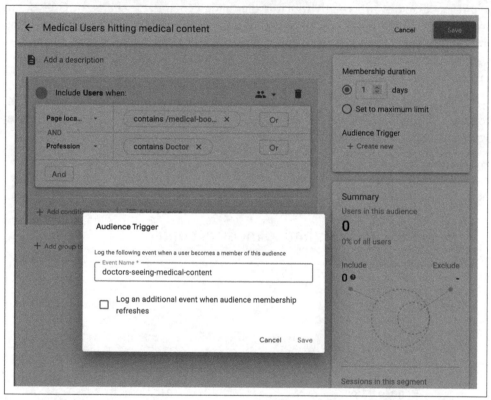

Figure 9-3. When users qualify for the audience, they can set off an event that can be seen within the Real-Time API

To fetch the data within R that will be the foundation of the dashboard, Shiny app is shown in Example 9-1.

Example 9-1. Fetching from the real-time API with the R library googleAnalyticsR

```
library(googleAnalyticsR)
ga_auth()
ga_id <- 1234567 # your GA4 propertyId

# fetch real-time audience data
```

```
ga_data(ga_id,
        metrics = "activeUsers",
        dimensions = c("minutesAgo", "audienceName"),
        realtime = TRUE)
#i 2022-06-17 12:47:40 > Realtime Report Request
#i 2022-06-17 12:47:41 > Downloaded [ 60 ] of total [ 60 ] rows
# A tibble: 60 × 3
#   minutesAgo audienceName                     activeUsers
#   <chr>      <chr>                            <dbl>
# 1 22         All Users                        335
# 2 13         All Users                        332
# 3 07         doctors-seeing-medical-content 29
# 4 09         doctors-seeing-medical-content 27
```

By this stage, we should have the GA4 events available that we want to query in real time via the Real-Time API. We now move to what we do with that data once we've captured and downloaded it.

Data Storage

Since we're using the GA4 Data API directly, GA4 itself will serve as the data storage for much of the data. The dashboard data will be small enough to fit within memory of the Shiny app displaying it, so there isn't much more to do regarding data storage other than deciding which server the Shiny app will be hosted on, which is covered in the next section.

Hosting the Shiny App on Cloud Run

There are many options for hosting Shiny apps, but for low numbers of users (< 10), I prefer to use the serverless options so I don't have to look after a server. You can do this on GCP using Cloud Run, a service much like Cloud Functions apart from the fact that it runs Docker images in the cloud. If the R Shiny app can be put into a Docker image, it can then be served from Cloud Run as any other HTTP website.

As an example, the Dockerfile shown in Example 9-2 could be used to create a Docker image that will run Shiny on Cloud Run.

Example 9-2. An example Dockerfile installing Shiny with googleAnalyticsR

```
FROM rocker/shiny

# install any R package dependencies
RUN apt-get update && apt-get install -y \
    libcurl4-openssl-dev libssl-dev

## Install extra packages from CRAN
RUN install2.r --error googleAnalyticsR
```

```
# copy over Shiny app into Shiny Server folder
COPY . /srv/shiny-server/

EXPOSE 8080

USER shiny

CMD ["/usr/bin/shiny-server"]
```

This Dockerfile can be placed in the same folder as your Shiny app (called `app.R` by default), along with any other configuration files you may need for your app such as a client for authentication purposes (`client.json`):

```
|
|- app.R
|- Dockerfile
|- client.json
```

To host this app on Cloud Run, we first need to create the Docker image within your own Google Project. You can do this via Cloud Build, which we covered in "Setting Up Cloud Build CI/CD with GitHub" on page 95 and which also has some Google documentation available for building containers (*https://oreil.ly/n6WEG*).

We need to host the Docker image so we can call it from Cloud Run; the image will be built and then stored at a location on GCP's Artifact Registry service (*https://oreil.ly/w6QWW*), a GCP service for hosting Docker images.

If we're building a Docker image, we don't need a *cloudbuild.yaml* file to configure the task, as it's a common enough job that if a Dockerfile is present, it knows what to do. In that case, we only need to specify the `--tag` flag to say where we want that Docker image to be pushed (the location on Artifact Registry). From the same folder, we submit the Cloud Build job via the gcloud command `gcloud builds submit --tag eu-docker.pkg.dev/learning-ga4/shiny/googleanalyticsr --timeout=20m`.

This should submit a Cloud Build job that will take quite a long time, which is why the timeout is increased via the `--timeout` flag.

Once it is built on Artifact Registry, you can then import that Docker image to run on Cloud Run, as can be seen in the setup screen shown in Figure 9-4.

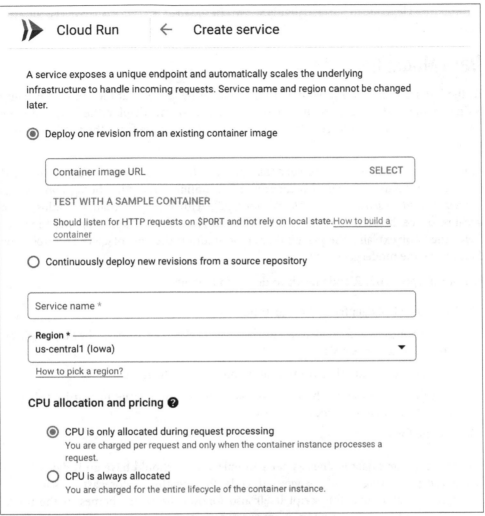

Figure 9-4. The container image URL will be the one you have specified for your local Docker build, such as `eu-docker.pkg.dev/learning-ga4/shiny/googleanalyticsr`

We're dealing with very "small" data here, since we don't need to keep a lot to make a dashboard—perhaps a thousand data points in total. Dashboards rarely need GBs of data because you only need the data points that a person can see, and there is only so much data one can digest. As such, the storage space for the instance running the dashboard doesn't need to be large, and the defaults are usually fine.

Where the work goes, though, is in what data is shown and how to turn your raw GA4 data into insight, which for this application will be a forecast within the Shiny dashboard. The code to surface that is shown in the next section, as we move into the data modeling phase.

Data Modeling

In the last section, we outlined how to host the Shiny server, but the reason for using Shiny in the first place is so it can run R code that can display the results in the browser. We now move to creating the forecast using R code from the GA4 real-time data.

For this forecast, I will use the forecast package (*https://oreil.ly/f6AEB*) maintained by Rob Hyndman, who has also coauthored an online forecasting book, *Forecasting: Principles and Practice* (*https://otexts.com/fpp2*), with George Athanasopoulos; it's a great resource. If you would like to deep-dive into how the R package creates the forecasts, the book explains the principles behind it and is the best place to research how to improve the model accuracy.

For our purposes our R code needs to do the following:

1. Input the GA4 data from the real-time API
2. Transform it to a format that can be used within the forecast library (e.g., an ordered time-series object)
3. Set seasonality and other configuration options for the forecast
4. Send the time-series to the forecast function to generate the forward prediction of points as well as a forecast interval
5. Plot the forecast

Assuming we have data fetches as per Example 9-1 we should have an R `data.frame` ready for processing. Some suggested code for creating a forecast is shown in Example 9-3. Note that the script itself also follows the main themes of the book, namely, collecting data, tidying it, modeling it, and then making an output for activation.

Example 9-3. Forecasting GA4 Real-Time data using `library(forecast)` and `library(googleAnalyticsR)`

```
library(googleAnalyticsR)
library(dplyr)
library(tidyr)
library(forecast)


# fetch real-time audience data
get_ga_rt <- function(ga_id){
```

```
  now <- Sys.time()

  rt_df <- ga_data(ga_id,
                   metrics = "activeUsers",
                   dimensions = c("minutesAgo",
                                  "audienceName"),
                   realtime = TRUE,
                   limit = 10000)
  # create a timestamp from (now - minutesAgo)
  rt_df$timestamp <- now -
      as.difftime(as.numeric(rt_df$minutesAgo),
                  units = "mins")

  rt_df

}

# tidy data by making each audience in its own column
tidy_rt <- function(my_df){

  my_df |>
    pivot_wider(names_from = "audienceName",
                values_from = "activeUsers",
                values_fill = 0) |>
    arrange(minutesAgo)
}

# forecast each audience activeUsers
forecast_rt <- function(rt){

  # remove non-forecastable columns
  rt$minutesAgo <- NULL
  rt$timestamp <- NULL

  # create time-series object
  rt_xts <- ts(rt, frequency = 60)

  # loop by column and make a list of forecasts for next 15mins
  forecasts <- lapply(rt_xts, function(x) forecast(x, h = 15))
  setNames(forecasts, names(rt))
}


# change to your GA4 property ID
ga_id <- 123456

# use functions above to make a list of forecasts per audience
forecasts <- get_ga_rt(ga_id) |> tidy_rt() |> forecast_rt()

# plot them to examine forecasts
lapply(forecasts, autoplot)
```

By default, the Real-Time API fetches only 30 minutes in the past, or 60 minutes on GA360. To enable our forecasts to work with more data, we'll keep the historic data from the API and append it to the latest fetch, so we can build up a reasonable history throughout the day. The code in Example 9-4 shows how you can do this to slowly build up data within your app for the forecasts. Again, we're only dealing with a low volume of data here even if we keep several hours' worth of data, so the Shiny app will easily handle the data volume (although you should probably cap it so it doesn't grow for weeks and weeks). In that case, we can store, say, up to 48 hours, which will represent `2880 rows * (number of Audiences)` amount of rows.

Example 9-4. Code to append your Real-Time API fetches so you can create a historic trend for your forecasts

```
library(googleAnalyticsR)
library(dplyr)
library(tidyr)

get_ga_rt <- function(ga_id){
  # fetch real-time audience data
  now <- Sys.time()
  rt_df <- ga_data(ga_id,
                   metrics = "activeUsers",
                   dimensions = c("minutesAgo", "audienceName"),
                   realtime = TRUE,
                   limit = 10000)
  rt_df$timestamp <- now - as.difftime(as.numeric(rt_df$minutesAgo), units = "mins")

  rt_df

}


tidy_rt <- function(my_df){

  my_df %>%
    pivot_wider(names_from = "audienceName",
                values_from = "activeUsers",
                values_fill = 0) %>%
    arrange(minutesAgo) |>
    filter(minutesAgo != "00") |>
    mutate(timemin = format(timestamp, format = "%d%H%M")) |>
    select(-minutesAgo)
}

append_df <- function(old, new){

  # do nothing if nothing to append
  if(is.null(old)) return(new)
  if(is.null(new)) return(old)
```

```
# rows that are in old but not in new
history <- anti_join(old, new, by = "timemin")

if(nrow(history) == 0) return(new)

# append the historic data, remove invalid minutesAgo
rbind(new, history) |>
    head(2880) # only keep top 48hrs (60*24*2)
}

# replace with your ga_id
ga_id <- 123456

# use like this
first_api <- get_ga_rt(ga_id) |> tidy_rt()

# wait longer than a minute then fetch again
second_api <- get_ga_rt(ga_id) |> tidy_rt()

# append rows from first_api that are not in second_api
append_df(first_api, second_api)

# repeat on schedule
```

We now have an R script that can process the real-time data and create some forecasts and regressions, but we need to activate this process so that the end users can react and make decisions based on this data, without needing to know R. For this, we need to create a dashboard using the data.

Data Activation—A Real-Time Dashboard

Let's look at data activation, taking the R scripts from the previous section and making a real-time dashboard out of the data for our colleagues. This will involve taking the standalone R script from "Data Modeling" on page 276 and hosting it within a Shiny server from "Data Storage" on page 273.

When we talk about "real-time," we have to consider exactly what that means. We said previously that a real-time dashboard needs real-time decisions to be of use, but how granular can these decisions be made? Does "real-time" mean responses within the second, or is every 10 minutes "real-time" enough?

The Real-Time API's granularity is by the minute, so practically speaking, a fetch every 60 seconds will be real-time enough unless you can think of reasons we need to react to data within subminute time frames. I cannot. This also factors into the quotas for the API, since a fetch every 10 seconds versus 1 minute will exhaust your quota 6 times faster for potentially no gain. Remember to check the API Quotas page (*https://oreil.ly/ueVEM*) to see how this sits with your own application.

In conclusion, this is justification for the fact that our "real-time" dashboard is actually going to have a response rate of 60 seconds, as we fetch data every minute to update the information on the dashboard.

R Code for the Real-Time Shiny App

The application we create in R could be replaced with your own in another dashboard system or via a web app in, say, Python, but I find it quickest within R.

To complement the modeled data coming from the R functions in the previous section, I will also reach for my favorite visualization JavaScript library, Highcharts (*https://www.highcharts.com*). This library allows you to make interactive visualizations that can be displayed in any browser. It also has a great R library for facilitating its use, including within Shiny, called `highcharter` (*https://oreil.ly/JfbSv*), written by Joshua Kunst. This library takes your R objects and turns them into the interactive JavaScript plots. A bit of interactivity is always good for your data presentation to allow some playing with the data by the end user. This will take the perfunctory plots we created before and add a bit of pizzazz to them. The code in Example 9-5 will take the forecast objects from Example 9-3 and turn them into a Highcharts version.

Example 9-5. Some example code that will take your raw and forecast data and turn them into highcharts plots

```
library(highcharter)

highcharter_plot <- function(raw,
                             forecast,
                             column = "All Users"){
  ## forcast values object
  fc <- forecast[[column]]

  ## original data
  raw_data <- ts(raw[,column], frequency = 60)

  raw_x_date <- as.numeric(raw$timestamp) * 1000

  ## create the right x-axis timestamps
  forecast_times <- as.numeric(
    seq(max(rt$timestamp),
        by=60,
        length.out = length(fc$mean))
        ) * 1000

  forecast_values <- as.numeric(fc$mean)

  # create the highcharts plot object
  highchart() |>
    hc_chart(zoomType = "x") |>
```

```
hc_xAxis(type = "datetime") |>
hc_yAxis(title = column) |>
hc_title(
  text = paste("Real-time forecast for", column)
) |>
hc_add_series(
  type = "line",
  name = "data",
  data = list_parse2(data.frame(date = raw_x_date,
                                value = raw_data))) |>
hc_add_series(
  type = "arearange",
  name = "80%",
  fillOpacity = 0.3,
  data = list_parse2(
    data.frame(date = forecast_times,
               upper = as.numeric(fc$upper[,1]),
               lower = as.numeric(fc$lower[,1])))) |>
hc_add_series(
  type = "arearange",
  name = "95%",
  fillOpacity = 0.3,
  data = list_parse2(
    data.frame(date = forecast_times,
               upper = as.numeric(fc$upper[,2]),
               lower = as.numeric(fc$lower[,2])))) |>
hc_add_series(
  type = "line",
  name = "forecast",
  data = list_parse2(
    data.frame(date = forecast_times,
               value = forecast_values)))
}
```

You can see an example of output in Figure 9-5.

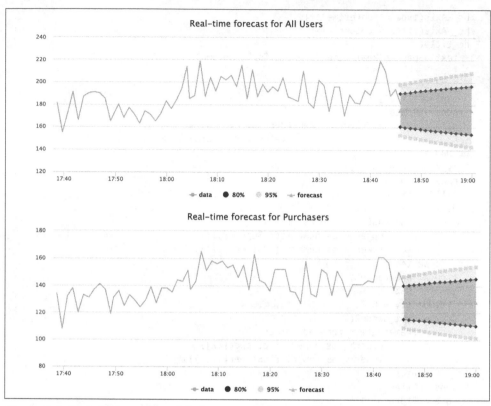

Figure 9-5. Output from the script that turns R forecast objects into `highcharts` *plots*

GA4 Authentication with a Service Account

The easiest and most secure way to enable access to your GA4 data within your app is to create a GCP service key and then give that key Viewer access to your GA4 data. You can then upload this service key with your app. This is not recommended, however, if your service key has access to any cloud resources that cost you money (like BigQuery) because if it's compromised, it could potentially cost you thousands of dollars.

You can create a service key within the Google Cloud Console (*https://oreil.ly/6N8Ur*) as shown in Figure 9-6. Do not assign any Cloud roles to the key. This one is called "fetch-ga" and will generate an email of the format *{name}*@*{project-id}*.iam.gser viceaccount.com, e.g., *fetch-ga@learning-ga4.iam.gserviceaccount.com*.

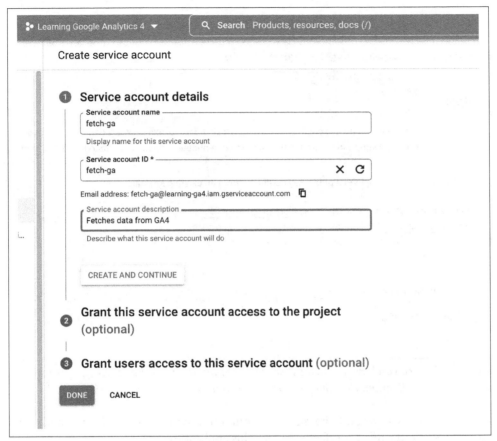

Figure 9-6. Create a service key for use within your app—if you give it no roles, then it can be used safely for GA4 without needing to worry that if it's compromised it will cost you cloud charges

The service email included with the JSON file can be treated exactly as you would treat your own or someone else's email who wants access to GA4, by adding it via the GA4 User Admin console. Before doing that, however, you will need a way for the app to authenticate itself, which is done via a JSON key associated with the service account. This is created in the same GCP console, as shown in Figure 9-7, which will generate a JSON file you can download to your computer. Keep this safe, and put it in a place where your application can reach it.

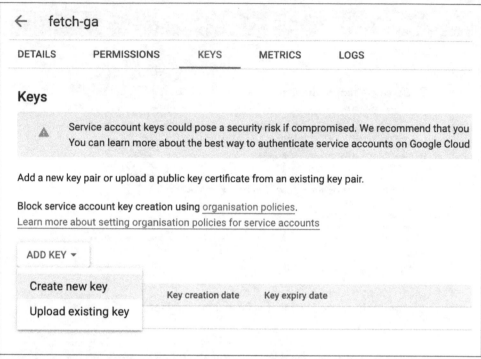

Figure 9-7. Once you have created the service account, download a JSON key for use within your application; this key gives access to your data, so guard it

All that remains is to add the email to your users within GA4 admin, as seen in Figure 9-8. Viewer role is perfectly fine for this application.

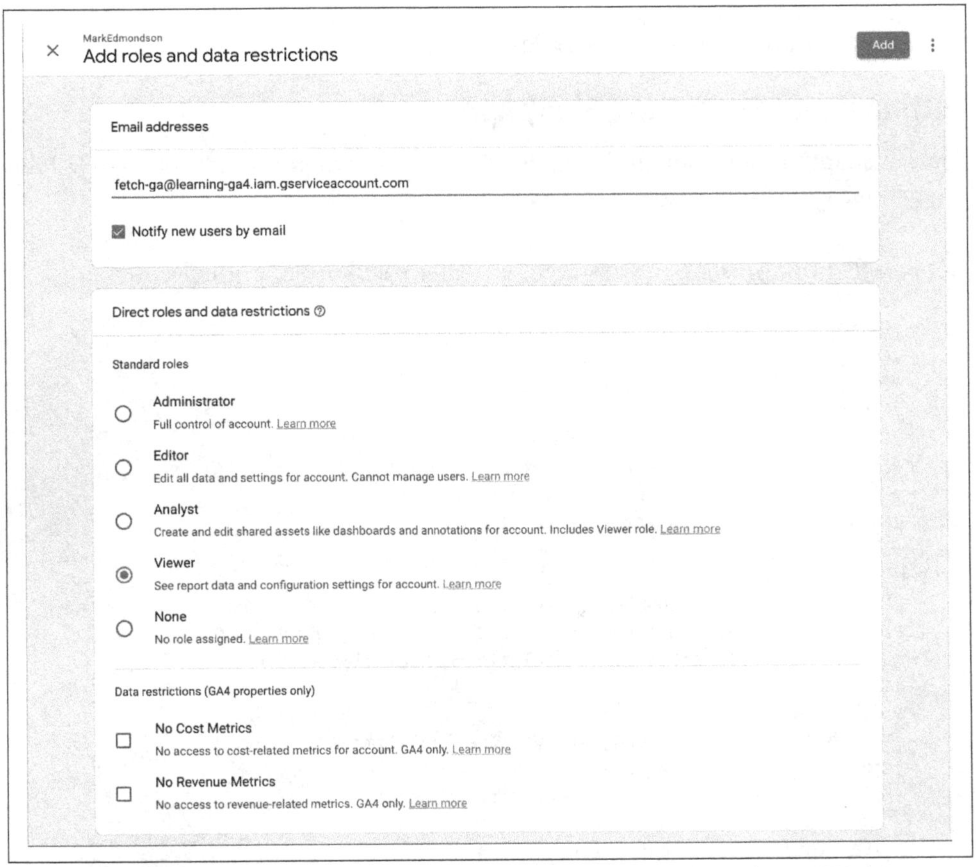

Figure 9-8. Adding a service email as a user to the GA4 interface for use within your scripts

This procedure will work for whatever language you're coding your GA4 API fetches for. For `googleAnalyticsR` in particular, you can then authenticate with this key by pointing the `ga_auth()` function at the file, as shown in Example 9-6.

Example 9-6. Authentication of GA4 using a JSON service key in `googleAnalyticsR`

```
library(googleAnalyticsR)

# authentication via json file set up to access GA4 account
ga_auth(json_file = "learning-ga4.json")
#> i Authenticating using fetch-ga@learning-ga4.iam.gserviceaccount.com

# test authentication by listing your accounts
ga_account_list("ga4")
#># A tibble: 2 × 4
#>   account_name  accountId property_name  propertyId
```

```
#>  <chr>           <chr>      <chr>           <chr>
#> 1 MarkEdmondson 47480439  GA4 Mark Blog  206670707
```

Putting It All Together in a Shiny App

You can adapt the Shiny app in Example 9-7 for your own purposes, pulling in all the
R functions within this chapter.

Example 9-7. Using the R functions from this chapter within a Shiny app

```r
library(shiny)            # R web apps
library(googleAnalyticsR) # getting GA4 data
library(tidyr)            # tidying data
library(forecast)         # modeling data
library(dplyr)            # data tidying
library(shinythemes)      # web app styling
library(DT)               # interactive html tables
library(highcharter)      # interactive plots

# the HTML UI of the app
ui <- fluidPage(theme = shinytheme("sandstone"),
    titlePanel(title=div(img(src="green-hand-small.png", width = 30),
              "Real-Time GA4"), windowTitle = "Real-Time GA4"),
    sidebarLayout(
      sidebarPanel(
        p("This app pulls in GA4 data via the Real-Time API using
        googleAnalyticsR::ga_data(),
        creates a forecast using forecast::forecast()
        and displays it in an interactive plot
        via highcharter::highcharts()"),
        textOutput("last_check")
      ),
      mainPanel(
        tabsetPanel(
          tabPanel("Realtime hits forecast",
                  highchartOutput("forecast_allusers"),
                  highchartOutput("forecast_purchasers"),

          ),
          tabPanel("Table",
                  dataTableOutput("table")
          )
        )
      )
    )
)
```

The functions from this chapter are all contained in the script and are collected in
Example 9-8.

Example 9-8. Functions for the Shiny application

```r
library(shiny)            # R web apps
library(googleAnalyticsR) # getting GA4 data
library(tidyr)            # tidying data
library(forecast)         # modeling data
library(dplyr)            # data tidying
library(shinythemes)      # web app syling
library(DT)               # interactive html tables
library(highcharter)      # interactive plots

get_ga_rt <- function(ga_id){
  # fetch real-time audience data
  now <- Sys.time()
  rt_df <- ga_data(ga_id,
                   metrics = "activeUsers",
                   dimensions = c("minutesAgo", "audienceName"),
                   realtime = TRUE,
                   limit = 10000)
  rt_df$timestamp <- now - as.difftime(as.numeric(rt_df$minutesAgo),
                                        units = "mins")

  rt_df

}

tidy_rt <- function(my_df){

  my_df |>
    pivot_wider(names_from = "audienceName",
                values_from = "activeUsers",
                values_fill = 0) |>
    arrange(desc(minutesAgo)) |>
    mutate(timemin = format(timestamp, format = "%d%H%M")) |>
    filter(minutesAgo != "00") |>
    select(-minutesAgo)

}

append_df <- function(old, new){

  if(is.null(old) || nrow(old) == 0) return(new)
  if(is.null(new) || nrow(new) == 0) return(old)

  # rows that are in old but not in new
  history <- anti_join(old, new, by = "timemin")

  if(nrow(history) == 0) return(new)

  # append the historic data
  rbind(history, new) |>
    head(2880) # only keep top 48hrs (60*24*2)
```

```
}

forecast_rt <- function(rt){

  rt$timestamp <- NULL
  rt$timemin <- NULL

  # ## the number of hits per timestamp
  rt_xts <- ts(rt, frequency = 60)

  do_forecast <- function(x, h = 30){
    tryCatch(
      forecast::forecast(x, h = h),
      error = function(e){
        warning("Could not forecast series - ", e$message)
      }
    )

  }

  forecasts <- lapply(rt_xts, do_forecast)

}

highcharter_plot <- function(rt, forecast, column = "All Users"){
  ## forcast values object
  fc <- forecast[[column]]

  ## original data
  raw_data <- ts(rt[,column], frequency = 60)

  raw_x_date <- as.numeric(rt$timestamp) * 1000

  ## each minute
  forecast_times <- as.numeric(
    seq(max(rt$timestamp), by=60, length.out = length(fc$mean))) * 1000

  forecast_values <- as.numeric(fc$mean)

  highchart() |>
    hc_chart(zoomType = "x") |>
    hc_xAxis(type = "datetime") |>
    hc_yAxis(title = column) |>
    hc_title(
      text = paste("Real-time forecast for", column)
    ) |>
    hc_add_series(
      type = "line",
      name = "data",
      data = list_parse2(data.frame(date = raw_x_date,
                                    value = raw_data))) |>
    hc_add_series(
```

```
      type = "arearange",
      name = "80%",
      fillOpacity = 0.3,
      data = list_parse2(
        data.frame(date = forecast_times,
                   upper = as.numeric(fc$upper[,1]),
                   lower = as.numeric(fc$lower[,1])))) |>
    hc_add_series(
      type = "arearange",
      name = "95%",
      fillOpacity = 0.3,
      data = list_parse2(
        data.frame(date = forecast_times,
                   upper = as.numeric(fc$upper[,2]),
                   lower = as.numeric(fc$lower[,2])))) |>
    hc_add_series(
      type = "line",
      name = "forecast",
      data = list_parse2(
        data.frame(date = forecast_times,
                   value = forecast_values)))
}
```

The server functions act as the backend for the Shiny app and populate the outputs as defined in Example 9-7. The backend functions call the functions as defined in Example 9-8. This is the code that runs the responsive R code, as shown in Example 9-9.

Example 9-9. The backend server functions for the application calling the functions to populate the data for the frontend UI

```
# Define server logic required to draw a histogram
server <- function(input, output, session) {

    # replace with your GA4 property ID
    ga_id <- 1234567

    historic_df <- reactiveVal(data.frame())

    get_ga_audience <- function(){
      # authentication via a JSON file
      ga_auth(json_file = "learning-ga4.json")

      # fetch data and tidy it
      get_ga_rt(ga_id) %>% tidy_rt()
    }

    # this is always different to force a real-time API call
    check_ga <- function(){
      Sys.time()
    }
```

```r
# checks every 31 seconds for changes
realtime_data <- reactivePoll(31000,
                              session,
                              checkFunc = check_ga,
                              valueFunc = get_ga_audience)

# builds up a historic record
historic_data <- reactive({
  req(realtime_data())

  # add new data to historic
  new_historic <- append_df(historic_df(),
                            realtime_data())

  # write new value to reactiveVal
  historic_df(new_historic)

  new_historic

})

# creates the forecast data objects
forecast_data <- reactive({
    req(historic_data())
    rt <- historic_data()
    message("forecast_data()")

    #
    forecast_rt(rt)
})

# outputs timestamp of last API call
output$last_check <- renderText({
  req(historic_data())

  last_update <- tail(historic_data()$timestamp, 1)

  paste("Last update: ", last_update)

  })

# a table of the raw data
output$table <- renderDataTable({
  req(historic_data())
  historic_data()
})

# create one of these per audience
## all users
output$forecast_allusers <- renderHighchart({
  req(forecast_data())
```

```
        highcharter_plot(historic_data(),
                         forecast_data(),
                         "All Users")

    })

    ## purchasers
    output$forecast_purchasers <- renderHighchart({
      req(forecast_data())

      highcharter_plot(historic_data(),
                       forecast_data(),
                       "Purchasers")

    })

    ## more audience plots here?

}

# Run the application
shinyApp(ui, server = server)
```

All things going well, you should see an app similar to that shown in Figure 9-9, updating every 60 seconds with the latest data.

Once it's working locally, you can deploy it as discussed in Example 9-2, with a Dockerfile created with the R packages and Shiny app and authentication JSON key all installed.

Routes to modify for your own purposes include picking the audience you want to track (such as the Medical Books Audience from the beginning of this chapter) and theming the app's look to better suit your organization.

Once you have a working dashboard, it's important to get feedback from the end users who you want to use it every day. They will no doubt give you feedback on things to improve, and getting feedback should be treated as an ongoing task to keep them happy and engaged with the data.

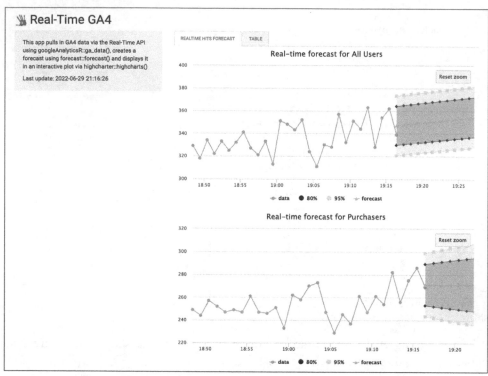

Figure 9-9. A running Shiny app with real-time GA4 data and a forecast (Highcharts visualization provided by `highcharter` *package calling* www.highcharts.com*)*

Summary

In this chapter, we discussed how to create a business case for your real-time dashboard, which resources you need to help scope out its capabilities, how to configure GA4 audiences so they are available in the real-time API, how to host a Shiny application within Cloud Run as the dashboard web app solution, how to properly configure authentication, how to create forecast modeling using R's `forecast` package, and how to put all the ingredients together for fetching data from GA4's real-time API, forecasting and presenting via Highcharts visualization libraries.

As with all the use cases presented, I don't know if the application will exactly map to your own needs, so I've tried to cover as many different scenarios and technologies I think are helpful so you can mix and match. As you build up experience, you will have more use cases under your belt, ideally with a new element each time. This way, you can build up your own catalog of techniques that you can apply to novel situations in the future. This book attempts to kick-start you with use cases I've seen in my career, but yours will be necessarily different. The most important lesson I hope you learn is to have a structured framework to be efficient in your learning. Google Cloud and GA4 are particularly helpful for this since there are always new innovations being released that you can apply the framework around, so keeping up-to-date with what is being released helps you keep on top of what is possible.

The techniques described within the book are ones I have personally gathered from many sources, and in the last chapter, we learn how to gather the necessary skills with some resources that have helped me over the years.

## CHAPTER 10

Next Steps

This chapter aims to leave you with resources you can use to build on the inspiration you've hopefully found in this book. This chapter also includes some resources you can use as a refresher, which may be useful if you found yourself out of your depth at all throughout the book.

As I've tried to emphasize, it's impossible to write a comprehensive book that covers every unique solution your business needs, but I hope you now have the tools to develop one yourself. I've written the use cases to cover many of the different aspects that have come up in my own history, but your business and experiences will be different. I've worked through many use cases, and each time I've aimed to introduce a new element, be it some code or service I've not tried before or a new approach. So when you're working on your GA4 project, maybe you can pick up this book and find an entry that will give you that extra element you didn't know before and can now include.

If you haven't implemented any of the solutions covered in the book, I encourage you to try at least one of them to help you experience for yourself any issues that may come up. Books and examples are idealized and don't cover the to-and-fro of fiddling with typos, forgotten links, or mistakes that I made when creating them.

In this last chapter, I'll share how I collected the knowledge to write this book, which is largely from the community and the kindness of strangers who have blogged, tweeted, or chatted in person about their own findings within the digital online communities.

Motivation: How I Learned What Is in This Book

A large part of education is motivation, so finding that motivation within yourself is a fundamental step for your future success. I'm not the right person to write about how to foster your own motivation (if you're reading this book, then I guess you're at least part way there already!), but I can tell you what motivated me with the hope it strikes a chord.

I enjoy creating new things and finding innovative solutions. In every project I work on, I try to include something new within it—even if it fails and I need to fall back to a tried-and-tested method, it's a valuable learning experience. It's a win-win each time since I then have either a better idea of what is best to use next time or I have a superior working solution than last time.

For me, I've also found that shifting my focus to a meta-level above what I'm working on every couple of years gives me a feeling of advancement. By this I mean working on the boundaries of what I know now so I can apply them more generally. For instance, I started in SEO working with websites, then got interested in how to measure SEO efforts and found my way into web analytics, then moved into how to scale web analytics solutions and found my way into the cloud. Working on the horizon of what I know now keeps my motivation high.

I find that another way to stay motivated is to by having an ideal solution in mind when you are working day-to-day. This ideal setup is one that implements all of your perfect features and assumes no resource, technical, or political issues (hard to imagine sometimes, I know!).

With such an ideal in mind and keeping it updated with the latest cutting-edge thinking, you will always have a goal to work toward. Some of the ideas in this book will help you realize some of these ideals. As you work toward your goal, you should enjoy the journey because while you may never reach it, you will be making useful stuff along the way.

My ideal setup would include the following:

- General business goals defined by the CEO reflected in the major KPIs measured within GA4 and across the business. Supporting website metrics for those KPIs would be understood and agreed upon by all stakeholders.

- A universal naming convention and schema available to everyone in the company, so everyone calls those metrics the same name and can relate them to the KPIs.

- A GTM dataLayer holding all necessary website metrics for all tags, with an agile marketing IT team ready to update as necessary.

- QA and testing of website releases including a pass/fail on the digital analytics data quality.

- A clear and ethical privacy policy for cookies and data governance for your user data, with maintaining those users' respect and trust regarded as a key goal. Ability for the user to reexamine what data they are passing to you and your third-party partners and revoke/opt-in easily.

- The ability for users to log-in once they do trust the website with one robust user ID that is generated centrally and used for all that customer's dealings with the business (cross channel, telephone, offline, etc.).

- Anonymous statistical tracking permitted for those who don'tt opt in to targeting.

- Ability for all digital marketers to self-service configurations and data points for their ad hoc analysis.

- Ability to trigger workflows based off any GA4 event sent in to activate data via your activation channels.

- Only the GA4 tag triggering on the website, sending all event data to a GTM SS implementation.

- All internal data systems on GCP, using BigQuery as the basis for their data warehouse, and other services such as Cloud Run, Cloud Functions, Cloud Build, etc., as needed.

- Enrichment of GA4 event streams with a Firestore lookup linked to backend systems.

- Audiences within GA4 enriched with backend data used to activate via Google Marketing Platform.

- Clean, tidy datasets within BigQuery holding the aggregations of both GA4 and internal systems, linked to the Looker BI tool for distribution within the business for ad hoc analysis.

- Tiered access to analytics systems: only analytics developers have access to the GA4 web console, reporting data delivered via the GA4 exports in BigQuery to Looker and/or Data Studio.

- Internal events regarding user activity fired through Pub/Sub, which can be optionally sent to GA4 to enhance analysis and be used in data activation use cases.

- A two-year development plan with a prioritized list of use cases with dependencies and resources all mapped out.

Yours may differ! But if I ever achieve a system that fulfills all of these requirements, then I may have to look for new challenges.

Learning Resources

The content of this book is learning from result of the many people I over the years, especially my many talented work colleagues, such as those at IIH Nordic in Copenhagen. Having a great team around while working for great people has helped enormously. I spend a large part of my day, though, consuming content from other members of the community online, and what follows is a curated list of some of the most helpful. If you haven't heard of some before, I encourage you to check them out:

Google development documentation
> The first may be the most obvious, but you'll be surprised by how many people don't actually read the documentation (*https://oreil.ly/27q8O*) that Google creates. Use the feedback button if you spot any errors or anything unclear, too, so it improves over time. I recommend you read everything at least once because it can clear up a lot of common questions and is always an authoritative link to send when answering queries.

Simo Ahava
> I worked with Simo Ahava before he was synonymous with GTM and digital strategy, so I have the privilege of knowing that he's just as helpful, trusted, and friendly as his public persona is. He deserves his success, and if you're looking to come to grips with GTM and his now-expanding interests with digital privacy and JavaScript fundamentals, sign up for his courses via his company, Simmer.
>
> - Simo's blog on mostly GTM and GA (*https://oreil.ly/FeKCh*)
> - Simmer online learning courses (*https://oreil.ly/0Zqjt*)
> - @SimoAhava (*https://oreil.ly/wyYsd*)

Krista Seiden
> Krista used to work for Google as a GA Evangelist, so she actually got to help shape GA itself. Now working independently and creating content to help with the transition to GA4, Krista has many resources that will help you get up to speed with using GA4, particularly when comparing it to Universal Analytics.
>
> - KS Digital, Krista's consultancy (*https://ksdigital.co*)
> - Krista's blog (*https://oreil.ly/kULC0*)
> - @kristaseiden (*https://oreil.ly/aLFGP*)

Charles Farina

Charles is often at the forefront of GA releases and regularly contributes either in the #measure Slack channel, on Twitter, or on his own blog. He is head of innovation at Adswerve, one the biggest GMP consultancies in the US.

- Charles Farina's blog (*https://oreil.ly/VO8Uc*)
- @CharlesFarina (*https://oreil.ly/WRXc4*)

Measure Slack

If you want to connect with more than 16,000 digital marketers, sign up to be a part of the #measure Slack community. It has dedicated channels for GA, data science, databases, and more and has become the largest digital marketing community online. It's also application-only (*https://www.measure.chat*), so the quality remains very good with a high signal-to-noise ratio.

Julius Fedorovicius

Julius is a relative newcomer to the digital analytics community but has quickly become a provider of tons of great information about how to get started with GA4, including video tutorials on his website, Analytics Mania (*https://oreil.ly/tHWXF*).

Ken Williams

Ken has been actively blogging (*https://ken-williams.com*) about key questions revolving around the transition from GA4 and Universal Analytics, and he writes both technical implementation guides and explanations on concepts such as how conversions are modeled.

Krisjan Oldekamp

Another relative newcomer to digital analytics publishing, Krisjan has provided some great tutorials at the intersection of Google Cloud and GA that this book also focuses on. On Stacktonic (*https://stacktonic.com*), Krisjan has a lot of Cloud integration posts that I wish I had written!

Matt Clarke

Matt Clarke's posts have been a long-term secret gem of mine with lots of tutorials on how to do practical data science using digital marketing data, including GA. He has also created a Python package for downloading GA4 data similar to my own R-based one.

- Practical Data Science blog (*https://oreil.ly/RpQJ4*)
- gapandas4, GA4 import library into Pandas (*https://oreil.ly/JE73r*)

Johan van de Werken

Johan was one of the first to write about how to work with the GA4 BigQuery exports, and this has blossomed into him offering a course on Simo's Simmer website. Johan's GA4BigQuery (*https://www.ga4bigquery.com*) website is a great resource that I still check when I've forgotten some syntax and has some SQL examples to quickly you get you up and running.

David Vallejo

David has quickly become the authority on how to work with and customize the actual JavaScript calls within your GTM and GA4 snippets, and extensive experience with GA4's Measurement Protocol. If you are looking at customization and advanced tracking setups, I would first check David's blog to see if he's done it already.

- David Vallejo's website (*https://oreil.ly/f6AcC*)
- GA4 Measurement Protocol Cheat Sheet (*https://oreil.ly/obEG2*)

"Churn Prediction for Game Developers Using Google Analytics 4 (GA4) and BigQuery ML"

This excellent YouTube tutorial (*https://oreil.ly/jK8Tv*) by Googlers Polong Lin and Minhaz Kazi shows how to combine GA4 data with BigQueryML, concentrating on reducing churn for a gaming app.

R for Data Science

If you want to get started with R, then I recommend *R for Data Science* (*https://r4ds.had.co.nz*) by Hadley Wickham and Garrett Grolemund (O'Reilly).

Forecasting: Principles and Practice

To get started with forecasting, I find this online book invaluable. It is by the author of the R forecast package, Rob J. Hyndman, and George Athanasopoulos. *Forecasting: Principles and Practice* (*https://otexts.com/fpp2*) contains examples using R but is also helpful as a general text.

The R and data science community

The community is always publishing various useful statistical techniques. A good place to start is the RWeekly.org (*https://rweekly.org*) newsletter. The data science blog *Towards Data Science* (*https://towardsdatascience.com*) includes various articles on statistical subjects.

This should give you plenty to keep you busy with learning resources. But even with all the resources in the world, you may need to ask for help, which we cover in the next section.

Asking for Help

I anticipate that you will need to self-diagnose issues that crop up even if you follow the examples to the letter. This is one of the most difficult barriers because knowing the right question to ask is almost as hard as finding the answers, and with less experience, identifying what exactly is wrong becomes more difficult. Asking the right question is as much of a skill as knowing the right answer. Here are some pointers that may help you in your quest:

- Read error messages and try to search for them online if they don't make sense to you. (This may sound flippant, but a surprising number of questions online are solved in the error message themselves.)
- Try to limit and test down to exactly the line or service that is causing you trouble. Commenting out random blocks of code is a valid tactic.
- Stack Overflow (*https://stackoverflow.com*) is a Q and A website that has saved me many times.
- If you're at a loss, add more logging to what you're doing. Print out the variables you're expecting, and see if they match your expectations.
- Set up a test regime early on in your process. Having test data you can always compare against can really speed up progress and is a worthwhile investment of time to save you frustration later.
- Understand the exact pipeline of what you are doing, and inspect each node to make sure it holds what you assume, e.g., the HTTP request from the browser, the data processed by a GA4 tag in GTM, the data in the debug view of GA4, etc.
- Intermittent errors are the hardest to track down because they probably have something to do with the environment or special circumstances of the request.

Once you've had some experience with your own projects and have learned from these resources, you may want to get certified so that you and others know you have hit a certain standard.

Certifications

Certifications can help signal to yourself and your employer that you are capable and have worked enough with your specialization to be recognized. If you're looking for a job, I think it's worth getting at least a few you can show to your potential employers. In that case, the cherry on top would be some demonstrations beyond the certificates that show you can apply what you learned to, say, an open source project.

There are many digital marketing certifications out there, but I've picked a few that I know are useful and I feel would help others:

- The GA4 training program (*https://oreil.ly/OkKbh*) has its own 50-question test, which will be the first benchmark that you know what you're talking about regarding GA4. All the material needed to pass it can be found within this book!
- Simo Ahava offers a certification process, "Simmer for Google Tag Manager" (*https://www.teamsimmer.com*).

- Krista Seiden offers GA4 courses (*http://academy.ksdigital.co*).

- GCP has many courses on Coursera, and the Professional Data Engineer (*https://oreil.ly/BikZr*) was useful when learning about all the relevant data services.

- The R Programming Coursera course (*https://oreil.ly/9fA3I*) helped me along my R journey.

These courses are useful because they cover the technologies I use in my everyday work, the tools of my trade. Once you pick the tools you want to specialize in, I encourage you to master them. Doing so gives me the deepest job satisfaction.

Final Thoughts

I wish you luck on your journey creating amazing GA4 integrations. It has been an interesting journey for me, and I've been lucky enough to also enjoy some public success. If I may pass on some of my luck and experience to you via content within this book so you can also benefit, then writing the book has been worthwhile. If there is one final thought I can leave you with, it's that I attribute a lot of my success to the decision to start publishing content in the community, both in blogging and via open source, to give back to the community that sustained me. It's shown me that we are never alone in the issues and problems we face, and sharing solutions has paid me back tenfold in the feedback and contributions from others. I look forward to any feedback and stories this book may spark for you, and if you wish, please do get in touch via my public channels.

Index

Surveys tool of Google Marketing Platform, 196

languages used in code, 13
predictive purchases, 19
    (see also predictive purchases use case)
real-time forecasting, 22
    (see also real-time forecasting use case)
serverless pyramid, 17
user data types, 41
  privacy by design, 42
user privacy (see privacy)
user properties, 10, 61-69
  Google Consent Mode, 62-68
  Google Signals, 68
  Measurement Protocol, 69-71
  no personalized ads, 61
  privacy respected, 10, 62
    predictive audiences, 239
  user identification methods, 68
    Reporting Identity, 68, 164
    user profession via userID field, 261
user-defined function (UDF)
  Dataflow to filter Pub/Sub topic fields, 148
user_id
  audience segmentation, 247
    (see also audience segmentation use case)
  customer relationship management (CRM), 247, 250
  events associated with user, 7
  key for linking datasets, 175
  Measurement Protocol, 69-71
  user identification methods, 68
    Reporting Identity, 164
  user properties, 10, 61-69
    no personalized ads, 61

V

Vallejo, David, 300
version control/Git, 43, 143
Vertex AI modeling for Google Cloud Platform, 183-185
  Docker container option, 188
  putting into production, 185

regional availability, 183
Video Intelligence API, 181
Vision API for OCR, 181
visualization of data
  about data activation, 203
  costs and data caching, 224
  dashboards
    aggregate tables, 223
    dashboards that work, 203, 268
    GA4 dashboarding options, 204-216
    GA4 Explorations, 210-216
    GA4 Reports, 205-210
    maybe not dashboards, 38, 203
  Data Studio, 217-220
    about, 195
  Highcharts JavaScript library, 280
    highcharter R library, 280-281
  Looker, 221-222
  R for data modeling, 187
  third-party tools, 222
VS Code IDE, 44

W

web and mobile app analytics unified, 2
Web Technologies Task View, 186
website hosting on Google Cloud Storage, 123
website testing with Google Optimize, 201-202
WebUI (Google Cloud Platform), 43
Werken, Johan van de, 78, 299
Wickham, Hadley, 102, 187, 300
Williams, Ken, 299
WRITE_TRUNCATE versus APPEND, 113

Y

YAML
  Cloud Functions Python code config.yaml file, 91
  Cloud Storage to BigQuery config file, 252
  cloudbuild.yaml
    about Cloud Build, 138
    Cloud Function redeployments, 98-100

About the Author

Mark Edmondson has worked in digital analytics for more than 15 years and is known as a contributor throughout the industry with his highly anticipated blogs and open source work that pushes the boundaries of what digital analytics can achieve. He is the author of several published R packages dealing with Google APIs, including googleAnalyticsR and googleCloudRunner, which he developed to help with his own work. After pursuing a masters in physics from King's College London, he has worked in all strands of digital marketing with world-wide brands up to his current interests in using the cloud, machine learning, and data science to turn data into information and insight. He is an international speaker on concepts such as machine learning, cloud computing, and data programming and is honored to be part of the Google Developer Expert program for Google Analytics and Google Cloud. Mark lives in Copenhagen, Denmark, with his wife, two children, cats, and guitars. If you'd like to get in touch, please do so via *ga4-book@markedmondson.me*.

Colophon

The animal on the cover of *Learning Google Analytics* is a white-footed sportive lemur (*Lepilemur leucopus*), also known as a dry-bush weasel lemur.

These medium-sized lemurs are endemic to southern Madagascar, primarily living in Didiereaceae forests full of spiny-leafed succulents or in tropical gallery rainforests that run along a rivers edge. Didiereaceae forests are known to have one of the driest and most unpredictable climates. While rainfall in these areas does occur, the soil drains moisture quickly and these areas are also prone to years of drought.

White-footed sportive lemurs have long limbs and tails that help them climb and leap between trees. The predominantly brown and grayish color of their fur helps them blend in with their habitat. Their backs, upper limbs, shoulders, and thighs are a dark brown. Their tails are a dark grayish brown and their bellies range from a pale gray to a creamy white. One of their most notable features is their large orange eyes that are outlined in black. Many people say they look like they are wearing a thick line of eyeliner because of this.

As nocturnal creatures, they forage for food at night and either sleep or defend their territory during the day. Because of the habitat they live in, they mostly eat a low-nutrient, leafy diet, which is sometimes supplemented with flowers and fruits when resources are scarce. Their lackluster diet means they have low energy levels, so they spend very little energy traveling between feeding areas.

Between the years 2000 and 2080, their population is expected to decline by 80%. The main threat to their survival is habitat loss due to burning practices that help develop livestock pasture and tree harvesting for charcoal production, timber, and planks.

White-footed sportive lemurs are specifically adapted to live in their habitat and do not survive in captivity, which greatly decreases their chance of survival as their forests shrink. Because of these factors, they are currently considered an endangered species. Many of the animals on O'Reilly covers are endangered; all of them are important to the world.

The cover illustration is by Karen Montgomery, based on an antique line engraving from *Mammalia*. The cover fonts are Gilroy Semibold and Guardian Sans. The text font is Adobe Minion Pro; the heading font is Adobe Myriad Condensed; and the code font is Dalton Maag's Ubuntu Mono.

O'REILLY®

Learn from experts.
Become one yourself.

Books | Live online courses
Instant Answers | Virtual events
Videos | Interactive learning

Get started at oreilly.com.